Universitext

Universitext

Editors (North America): J.H. Ewing, F.W. Gehring, and P.R. Halmos

Soo Bong Chae

Lebesgue Integration

Second Edition

With 25 Illustrations

Springer-Verlag
New York Berlin Heidelberg London Paris
Tokyo Hong Kong Barcelona Budapest

Soo Bong Chae
(deceased)

Proofreading of this volume was done by Tony G. Horowitz.

Mathematics Subject Classifications (1991): 28-01, 28A25

Library of Congress Cataloging-in-Publication Data
Chae, Soo Bong, 1939–1994
    Lebesgue integration / Soo Bong Chae.—2nd ed.
      p. cm.
    Includes bibliographical references and index.
    ISBN 0-387-94357-9 (New York : acid-free). — ISBN 3-540-94357-9
(Berlin : acid-free)
    1. Integrals, Generalized. I. Title.
    QA312.C47 1995
    515′.43—dc20   94-27962

Printed on acid-free paper.

Production coordinated by Brian Howe and managed by Bill Imbornoni; manufacturing supervised by Genieve Shaw.
Typeset by Asco Trade Typesetting Ltd., Hong Kong.
Printed and bound by R.R. Donnelley & Sons, Harrisonburg, VA.
Printed in the United States of America.

9 8 7 6 5 4 3 2 1

ISBN 0-387-94357-9 Springer-Verlag New York Berlin Heidelberg
ISBN 3-540-94357-9 Springer-Verlag Berlin Heidelberg New York

Preface to the Second Edition

Responses from colleagues and students concerning the first edition indicate that the text still answers a pedagogical need which is not addressed by other texts.

There are no major changes in this edition. Several proofs have been tightened, and the exposition has been modified in minor ways for improved clarity.

As before, the strength of the text lies in presenting the student with the difficulties which led to the development of the theory and, whenever possible, giving the student the tools to overcome those difficulties for himself or herself.

Another proverb:

> Give me a fish, I eat for a day.
> Teach me to fish, I eat for a lifetime.

March 1994   Soo Bong Chae

Preface to the First Edition

This book was developed from lectures in a course at New College and should be accessible to advanced undergraduate and beginning graduate students. The prerequisites are an understanding of introductory calculus and the ability to comprehend "$\varepsilon-\delta$ arguments."

The study of abstract measure and integration theory has been in vogue for more than two decades in American universities since the publication of *Measure Theory* by P.R. Halmos (1950). There are, however, very few elementary texts from which the interested reader with a calculus background can learn the underlying theory in a form that immediately lends itself to an understanding of the subject. This book is meant to be on a level between calculus and abstract integration theory for students of mathematics and physics.

There is *much* time and opportunity to become abstract in a program of mathematics. We present the Lebesgue theory in a manner which gives beginners a sufficient base of examples and renders the abstract ideas credible and natural.

Although a familiarity with topological concepts on the real line is required of the reader, we begin with Chapter Zero, which can be a good review of the subject. The advanced reader may skip this chapter.

We present the Riemann integral in Chapter I to show the reader the necessity of a new concept of integration. In particular, we introduce the concept of sets of measure zero and characterize Riemann integrable functions in terms of this. This concept plays an intrinsic role in the subsequent chapters.

There are many ways to present Lebesgue's theory. Roughly, the various approaches fall into two main categories. In the first category are those in which measure comes first and integration comes second, such as in Lebesgue's dissertation. In the other category the order is reversed. Our approach is in the second. Since Lebesgue's thesis, many essentially equivalent approaches have been found.

In Chapter II we introduce the Lebesgue integral on a closed interval starting with the elementary concept of the integral of a step function à la F. Riesz. This approach avoids the assumption of any knowledge of the Riemann theory. Another advantage of this method is that it makes it possible to prove at the outset (and on the basis of only rudimentary knowledge of sets of real numbers) the fundamental theorems of Lebesgue which state the precise conditions under which term-by-term integration is allowed. In Chapter III, the theory of measure follows from the theory of integration.

Chapter IV is devoted to generalizations of the integral concept on a closed interval to more general sets. In this chapter we also study multiple integrals and their relation to iterated integrals.

In Chapter V, we study in depth the fundamental theorem of calculus for the Lebesgue integral. In doing so we use F. Riesz's Rising Sun Lemma and L.A. Rubel's proof of differentiability of monotone functions.

In order to whet the reader's appetite for functional analysis, we present the L^p spaces in Chapter VI as an application of the Lebesgue theory. This topic then leads to Hilbert spaces and the L^2 theory of Fourier series. Chapter VI is independent of Chapter V.

We often forget that it took many brilliant men numerous years to develop what we study in one term. I have tried to inform the reader of some historical facts about the subject, but I fear that I also have followed "Boyer's Law": *Mathematical formulas and theorems are usually not named after their original discoverers* (see Kennedy, 1972). For a further study of the history of the subject, we refer to Hawkins (1970). Lebesgue's address given at a conference in Copenhagen on May 8, 1926 to the Society for Mathematics is appended to give the reader an opportunity to know the originator's own thoughts. This translation attempts to maintain his humor without revisions or modifications.

This book can be used in several ways: it can be used as a textbook for a course in real analysis, for independent study, or as a supplementary text for a course in abstract integration theory.

In teaching the course, I emphasized as much as I could a firm belief that the joy of mathematics is in doing, not in hearing or seeing it.

> I hear, and I forget;
> I see, and I remember;
> I do, and I understand.
>
> (an old oriental proverb)

In class, I note the important points of each upcoming section, explain key definitions, theorems, and sketches of proofs, and indicate what the problems are about, before assigning the section and the problems I wish to cover. I can then let the text do much of what a formal lecture might do for that section.

Although the physical labor of organizing and writing this book was mine alone, it is obvious that I am deeply indebted to all of the mathematicians who have taught, guided, and inspired me. More personally, I wish to express my thanks to three teachers of undergraduate and graduate years: Professors Kyung Whan Kwun, Henry Sharp, Jr., and Leopoldo Nachbin. I am grateful to my colleague, Professor William K. Smith, without whose encouragement and counsel I would never have written this book. Special acknowledgments must go to New College students over the past several years, who read the manuscript with care and made numerous helpful suggestions and who let me know that they enjoyed the course—especially Tom Peters, George Konstantinow, Dr. Bonnie Saunders, Dr. Don Goldberg, Dr. Vincent Peck, Dr. John Smillie, James Foster, Robert Gayvert, and many others. I wish to thank them all. George Konstantinow translated Lebesgue's address, and Rob Gayvert professionally converted the final manuscript into complete camera-ready copy in accordance with the publisher's demands, a Herculean task. The illustrations were drawn by Jean Angelos. I would like to thank New College for encouragement and support; Provost Eugene Lewis especially has been most helpful. A final and special thanks must also go to Sookkyung, Dusan, and Nabin for bearing with me through all the hours spent on this rather than with them.

Soo Bong Chae

Contents

Contents

CHAPTER ZERO
Preliminaries

The purpose of this chapter is not to serve as a text on set theory, the real number system, and topology, but to indicate to the beginner exactly which concepts and results to familiarize oneself with before studying Lebesgue integration. To save the reader unnecessary effort, we shall develop most of the topics at as elementary a level as possible.

§1. Sets

In this section we shall describe some notions from set theory which will be useful. Our purpose is descriptive, and the arguments given are directed toward plausibility rather than toward rigorous proof. Naturally, we shall adopt a naive viewpoint in developing an elementary theory of sets. The great German mathematician Georg Cantor (1845–1918) is regarded as the creator of the theory of sets. For a further study of the topic we refer the interested reader to Halmos (1960).

A *set* is to be thought of intuitively as a collection of objects. This is not a definition of a set because the word *collection* is only synonymous with the word *set*. No attempt will be made to define these terms or to present a list of axioms for set theory. In order to avoid certain paradoxes that might arise from forming new sets from old ones, we shall have some set X fixed for a given discussion and consider only sets whose elements are subsets of X and so forth. Unfortunately, if we do not restrict formation of new sets from old ones, we have the so-called Russell's paradox:

Let U be the set of all sets that do not belong to themselves.

Does U belong to itself?

Throughout this book whenever the word *set* is used, it will be interpreted as a subset of a given set X.

Sets will be denoted by capital letters, such as A, B, ..., and elements (or members) of sets by lowercase letters, such as a, b, ... The set with elements a, b, c, ... is often denoted by

$$\{a, b, c, \ldots\}.$$

In what follows we shall take for granted the following sets, which occur throughout mathematics:

$\mathbb{N} = \{1, 2, 3, \ldots\}$, the set of all natural numbers;
$\mathbb{Z} = \{0, 1, -1, 2, -2, 3, -3, \ldots\}$, the set of all integers;
$\mathbb{Q}$, the set of all rational numbers;
$\mathbb{R}$, the set of all real numbers; and
$\mathbb{C}$, the set of all complex numbers.

If a is an element of A, the notation

$$a \in A$$

means that "*a belongs to A*." The negation of this assertion, "*a does not belong to A*," will be denoted by

$$a \notin A.$$

Thus, for example, for every element a of A, we have $a \in A$, and for no element a of A do we have $a \notin A$. If every element of A belongs to B, we say that A is a *subset* of the set B and write

$$A \subset B \qquad \text{or} \qquad B \supset A.$$

Two sets A and B are called *equal* and written $A = B$ if and only if they consist of precisely the same elements; or, equivalently, if and only if

$$A \subset B \qquad \text{and} \qquad B \subset A.$$

If $A \subset B$ but $A \neq B$, we call A a *proper subset* of B.

It is convenient to introduce the concept of the *empty set* for simplification in language and notation. A set is said to be *empty* if and only if it has no elements. The empty set will be denoted by $\varnothing$. For every set A we have

$$\varnothing \subset A.$$

For every object a we have

$$a \notin \varnothing.$$

The subsets of a given set X are frequently defined by imposing conditions upon the elements of X. For example, if $P(x)$ denotes a given statement relating to the element x of X, then

$$\{x \in X : P(x)\}$$

denotes the set of those elements x for which the proposition $P(x)$ is true. For example, $\mathbb{Q} = \{m/n : m, n \in \mathbb{Z}, \text{ and } n \neq 0\}$.

There are many ways of forming sets from old ones. The following two operations are fundamental. Given sets A, B we may form two sets from them:

$$A \cup B = \{x: x \in A \text{ or } x \in B\},$$

$$A \cap B = \{x: x \in A \text{ and } x \in B\}.$$

We call $A \cup B$ the *union* and $A \cap B$ the *intersection* of A and B.

We shall often want to form the union or intersection of a collection (or class) of sets. Let $\mathscr{C}$ be a collection of sets A. Then we define

$$\bigcup \{A: A \in \mathscr{C}\} = \{x: x \in A \text{ for some } A \in \mathscr{C}\},$$

$$\bigcap \{A: A \in \mathscr{C}\} = \{x: x \in A \text{ for all } A \in \mathscr{C}\}.$$

Sometimes it is convenient to write

$$\bigcup A_\alpha, \quad \bigcap A_\alpha,$$

where we regard α as running through some indexing set. If α runs through $\mathbb{N}$ we usually write

$$\bigcup \{A_n: n \in \mathbb{N}\} = \bigcup_{n=1}^{\infty} A_n$$

and similarly for $\bigcap_{n=1}^{\infty} A_n$. It is emphasized that $\infty \notin \mathbb{N}$, and hence there is no A_∞ in the collection. The "∞" in this notation is merely conventional. The *difference* between A and B, denoted by $A \backslash B$, is defined to be the set

$$A \backslash B = \{x: x \in A \text{ and } x \notin B\}.$$

If A is a subset of a given set X, the difference $X \backslash A$ will be called the *complement* of A with respect to X. If we consider only subsets of a fixed set X, we denote $X \backslash A$ by $\mathsf{C} A$. It is clear that $A \backslash B = A \cap \mathsf{C} B$, $\mathsf{CC} A = A$, and that $A \subset B$ is equivalent to $\mathsf{C} B \subset \mathsf{C} A$. The two following results concerning complementation are known as *De Morgan's laws* after Augustus De Morgan (1806–1871):

$$\mathsf{C}\left(\bigcup A_\alpha\right) = \bigcap (\mathsf{C} A_\alpha),$$

$$\mathsf{C}\left(\bigcap A_\alpha\right) = \bigcup (\mathsf{C} A_\alpha).$$

The following properties of union and intersection are easy to show:

(a) $\bigcap A_\alpha \subset A_\beta \subset \bigcup A_\alpha$ for any β;
(b) $A \cup (\bigcap A_\alpha) = \bigcap (A \cup A_\alpha)$; and
(c) $A \cap (\bigcup A_\alpha) = \bigcup (A \cap A_\alpha).$

§2. Relations

Let a, b be any objects. Then the *ordered pair* (a, b) is defined as

$$(a, b) = \{\{a\}, \{a, b\}\}.$$

It can easily be shown that

$$(a, b) = (a', b') \qquad \text{if and only if} \quad a = a' \text{ and } b = b'.$$

This is the crucial property of the ordered pair. Any other construction with this property could be used instead. Notice that (a, b) is quite different from $\{a, b\}$, since $\{a, b\}$ is always equal to $\{b, a\}$.

The *Cartesian product* of sets A and B, written $A \times B$, is the set of all ordered pairs (a, b) such that a belongs to A and b belongs to B, i.e.,

$$A \times B = \{(a, b): a \in A \text{ and } b \in B\}.$$

If $A = \{1, 2, 3\}$ and $B = \{a, b\}$, then the Cartesian product is the set

$$A \times B = \{(1, a), (1, b), (2, a), (2, b), (3, a), (3, b)\}.$$

The Cartesian product $\mathbb{R} \times \mathbb{R}$ of the real line with itself is the Euclidean plane $\mathbb{R}^2$ [hence the name Cartesian, after the French mathematician and philosopher René Descartes (1596–1650), who created plane analytic geometry]. Inductively, we can define $\mathbb{R}^n = \mathbb{R} \times \cdots \times \mathbb{R}$ (n times).

A *relation between sets* A and B is a subset R of $A \times B$. Examples of such an R are

$$\{(1, a), (1, b)\}, \qquad \{(2, a), (3, b)\}, \qquad \text{and} \qquad \{(1, a), (3, a)\}.$$

as taken from the above example. Two trivial relations from A to B are the sets $\varnothing$ and $A \times B$.

If R is a relation between sets A and B, then the fact that an element $a \in A$ bears the relation R to $b \in B$ may be expressed in the form $(a, b) \in R$ or, as is more commonly written, aRb.

A relation $R \subset A \times A$ is called an *equivalence relation on* A if it is:

(a) Reflexive: aRa for all $a \in A$.
(b) Symmetric: aRb implies bRa.
(c) Transitive: aRb and bRc imply aRc.

Equality is obviously an equivalence relation on any set. Conversely, an equivalence relation can always be replaced by the equality relation between suitable sets. In fact, let R be an equivalence relation on A. For a given $a \in A$, let $[a]$ be the set of elements equivalent to a, i.e., $[a] = \{b : aRb\}$. It is clear that $a \in [a]$. The set $[a]$ is called the *equivalence class* containing the element a.

2.1. Proposition. *Let R be an equivalence relation on A. Then:*

(1) *aRb if and only if* $[a] = [b]$;
(2) $a \in [a]$; *and*
(3) $[a] \cap [b] \neq \emptyset$ *implies* $[a] = [b]$.

The theorem means that the equivalence classes $[a]$ divide the set A in a manner such that A is the disjoint union of the equivalence classes under R.

Proof. (1) Suppose aRb. If $c \in [a]$, then cRa and, by transitivity, cRb, so that $c \in [b]$. Thus $[a] \subset [b]$. In a similar argument, $[b] \subset [a]$; it follows that $[a] = [b]$.

Next assume that $[a] = [b]$. Since $a \in [a]$, it follows that $a \in [b]$; hence aRb.

(3) Suppose that $c \in [a] \cap [b]$. Then cRa and cRb. Hence aRb, and $[a] = [b]$ follows from (1). □

The most significant type of relation that occurs in mathematics is that which is called a function. The following definition of a function may seem rather strange to those who are used to calculus and analysis texts which extensively employ functions but never actually define them.

A *function* f from A into B is a relation between A and B such that, for each $a \in A$, there is exactly one $b \in B$ such that $(a, b) \in f$. We write $f(a) = b$ to mean $(a, b) \in f$. Other terms for function are *mapping* and *transformation*.

Our concept of a function as a certain set of ordered pairs is what some would call the *graph of a function*, since an elementary definition of a function is rather a rule or something similar. We shall use the term "graph of a function" when this seems more expressive.

Let us return to general relations. The *domain* of a relation is the set of all first coordinates of its elements. The *range* is the set of all second coordinates. The notation

$$f : A \to B$$

is interpreted as "f is a function from the set A into the set B such that A is the domain of f and the range of f is a subset of B, not necessarily the whole of B." For example, define f by $f(x) = e^x$, for $x \in \mathbb{R}$. Then the domain of f is $\mathbb{R}$ and the range of f is the set

$$\mathbb{R}^+ = \{x \in \mathbb{R} : x > 0\}.$$

We may write, with increasing accuracy,

$$f : \mathbb{R} \to \mathbb{R} \quad \text{and} \quad f : \mathbb{R} \to \mathbb{R}^+.$$

A function f is called a *mapping from A onto B* if, for each $b \in B$, there is *at least one* $a \in A$ such that $b = f(a)$.

A function f is called a *one–one mapping* from A to B if, whenever $a, a' \in A$ and $a \neq a'$, then $f(a) \neq f(a')$. In other words, f is a one–one mapping if the two relations $f(a) = b$ and $f(a') = b$ imply that $a = a'$.

The function $f: \mathbb{R} \to \mathbb{R}^+$ defined by $f(x) = e^x$ is one–one and onto.

Exercise 2

A. Let A and B be sets and A^B denote the set of all functions from B to A. Show that

$$A^{\varnothing} = \{\varnothing\} \quad \text{and} \quad \varnothing^B = \varnothing \quad \text{if } B \neq \varnothing.$$

§3. Countable Sets

Two sets A and B are said to be *equivalent*, in symbols $A \sim B$, if there is a one–one mapping from A onto B. We note that $A \sim A$, $A \sim B$ implies $B \sim A$, $A \sim B$ and $B \sim C$ imply $A \sim C$. Hence $\sim$ is an equivalence relation.

If $A = \varnothing$ or $A \sim \{1, 2, \ldots, n\}$ for some $n \in \mathbb{N}$, then A is called *finite*. If A is finite or equivalent to $\mathbb{N}$, then A is called *countable*. Otherwise it is called *uncountable*. Obviously $\mathbb{N}$ is countable.

We can write the elements of a nonempty countable set in the form

$$\{a_1, a_2, a_3, \ldots\},$$

where a typical element of the set would be denoted by a and the subscripts are the consecutive natural numbers which indicate a one–one correspondence between the set and $\mathbb{N}$.

We are now in a position to prove some simple propositions about countable sets. It is clear that every subset of a countable set is countable.

3.1. Proposition. *The union of a countable number of countable sets* $A_1, A_2, \ldots$ *is itself countable.*

Proof. We may assume that $A_1, A_2, \ldots$ are mutually disjoint, that is, $A_m \cap A_n = \varnothing$ if $m \neq n$. Otherwise we could consider the sets $A_1, A_2 \backslash A_1, A_3 \backslash (A_1 \cup A_2)$, $\ldots$, instead. If we write, for each n,

$$A_n = \{a_{n1}, a_{n2}, a_{n3}, \ldots\},$$

then we can count all the elements in the union of the A_n's one by one in the manner indicated in the following table:

$$
\begin{array}{cccccc}
a_{11} & a_{12} & a_{13} & a_{14} & \cdots \\
a_{21} & a_{22} & a_{23} & a_{24} & \cdots \\
a_{31} & a_{32} & a_{33} & a_{34} & \cdots \\
a_{41} & a_{42} & a_{43} & a_{44} & \cdots \\
\vdots & \vdots & \vdots & \vdots
\end{array}
$$

It is clear that this procedure associates a unique number to each element in the union, hence establishing a one–one correspondence between the union and N. □

3.2. Proposition. *The set Q of all rational numbers is countable.*

Proof. The set $A_n = \{m/n: m \in \mathbb{Z}\}$ is countable for each $n \in \mathbb{N}$ since $\mathbb{Z}$ is countable. Since $\mathbb{Q} = \bigcup_{n=1}^{\infty} A_n$, it follows from Proposition 3.1 that $\mathbb{Q}$ is countable. □

EXERCISES 3

A. Show that a subset of a countable set is countable.

B. Prove that the collection of all finite sets of N is countable.

C. Let $\omega = \{0, 1, 2, 3, \ldots\}$. Define $f: \omega \times \omega \to \omega$ by

$$
f(m, n) = n + \tfrac{1}{2}k(k + 1) \qquad \text{where} \quad k = m + n.
$$

Show that f is one–one and onto. The following table depicts this relation:

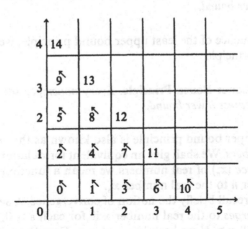

§4. Real Numbers

Since we have to start somewhere, we shall assume the reader to be familiar
with the real number system. Its algebraic properties related to addition,
subtraction, multiplication, division, and inequalities will be taken completely for granted. We shall discuss here the most crucial fundamental property of real numbers on which the whole development of real analysis, and
hence a great part of all analysis, depends. This property can be stated in
many equivalent forms, and the particular one that we select is a matter of
taste.

Before we state the fundamental so-called least upper bound principle, we
need some more terminology. Let E be a nonempty set of real numbers. We
say that E is *bounded above* if there is a number b such that $x \le b$ for all $x \in E$.

Then b is called an *upper bound* of E. If E is bounded above, then an upper
bound c is called a *least upper bound* (or *supremum*) of E if it is less than or
equal to any other upper bound of E. When this number exists, we denote it
by $c = \sup E$. A *greatest lower bound* (or *infimum*), denoted by $\inf E$, is defined similarly. Clearly, $\sup E$ and $\inf E$ are unique if they exist.

When we say that a set has a least upper bound or a greatest lower bound,
we are making no statement as to whether the set contains them as elements
or not.

Since a least upper bound of a set E is a special upper bound, it is clear
that only sets bounded above can have a least upper bound. However, the
empty set $\varnothing$ has no least upper bound even though it is bounded above by
any real number (see Exercise 4B). Therefore it is plausible to state the following fundamental principle, which we take as one of the axioms about real
numbers.

4.1. Least Upper Bound Principle. *Every nonempty set E that is bounded above
has a least upper bound.*

As a consequence of the least upper bound principle, we have the greatest
lower bound principle.

4.2. Greatest Lower Bound Principle. *Every nonempty set E that is bounded
below has a greatest lower bound.*

The least upper bound principle is also known as the *completeness axiom
of the real numbers*. We shall give an equivalent formulation of this principle.

By a *sequence* (x_n) of real numbers we mean a function which maps each
natural number n to the real number x_n.

Let us now recall briefly the notion of convergence. A *sequence* (x_n) of real
numbers *converges* to the real number x if, for each $\varepsilon > 0$, there is an $N > 0$
such that for all $n > N$ we have $|x_n - x| < \varepsilon$. Then the number x is called a

limit of the sequence (x_n) and we write

$$x = \lim_{n \to \infty} x_n, \qquad x = \lim_n x_n, \qquad \text{or} \qquad x_n \to x.$$

It is clear that a sequence can have at most one limit.

A sequence (x_n) of real numbers is called a *Cauchy sequence* if, given $\varepsilon > 0$, there is an N such that for all $n > N$ and all $m > N$ we have $|x_n - x_m| < \varepsilon$. In calculus it is shown that *every convergent sequence is a Cauchy sequence and every Cauchy sequence converges to exactly one number.* The latter property is usually referred to as the *completeness* of the real number system. The least upper bound principle is then equivalent to the completeness of the real number system (Exercise 4J). For this reason, the least upper bound principle is called the *completeness axiom*.

Sometimes it is convenient to use the following notations: If E has no upper bound, we write sup $E = \infty$. Likewise, if E has no lower bound, we write inf $E = -\infty$. It is plausible to write sup $\varnothing = -\infty$ and inf $\varnothing = \infty$ (why?).

If a set E of real numbers is bounded both above and below, it is called *bounded*. A bounded nonempty set E is characterized by having both a finite sup E and a finite inf E, or equivalently, by being contained in some finite interval (a, b). We shall study bounded sets in greater depth in the next two sections.

Despite the fact that the set of rational numbers is countable (see §3), we have the following proposition:

4.3. Proposition. *The set of all real numbers is uncountable.*

Proof. To demonstrate this, it suffices to prove that the interval $[0, 1]$ is uncountable (why?). We shall use the elegant argument of Georg Cantor. We use the technique called *Cantor's diagonal process*. We assume it is known that every real number x in $[0, 1]$ has a decimal representation of the form

$$x = 0.a_1 a_2 a_3 \ldots,$$

where each a_k is an integer between 0 and 9 inclusive.

Let $\{x_1, x_2, x_3, \ldots\}$ be a countable subset of $[0, 1]$, and let

$$x_1 = 0.a_{11} a_{12} a_{13} \ldots,$$

$$x_2 = 0.a_{21} a_{22} a_{23} \ldots,$$

$$x_3 = 0.a_{31} a_{32} a_{33} \ldots,$$

$$\cdots\cdots\cdots\cdots\cdots\cdots\cdots$$

Now set $a_n = 5$ if a_{nn} is even, and $a_n = 8$ if a_{nn} is odd. Consider the number y with the decimal representation

$$y = 0.a_1 a_2 a_3 \ldots.$$

Evidently y satisfies $0 < y < 1$ and $y \neq x_m$ for all m. Therefore, any countable subset of [0, 1] will omit at least one real number in this interval. This shows that every countable subset of [0,1] is a proper subset of [0, 1]. Therefore [0, 1] is not countable.   □

Exercises 4

A. Show that a nonempty finite set of real numbers has a least upper bound and a greatest lower bound.

B. Show that the empty set is bounded by any real number.

C. Prove that if a subset E of real numbers contains an upper bound, then this upper bound is the least upper bound of E.

D. Give an example of a set of rational numbers which is bounded above but which does not have a rational least upper bound.

E. Give a proof of the greatest lower bound principle (4.2).

F. Show that a sequence can have at most one limit.

G. Show that each Cauchy sequence is bounded.

H. Show that if a Cauchy sequence has a subsequence which converges to x, then the original sequence converges to x.

I. Show that a sequence of real numbers converges if and only if it is a Cauchy sequence.

J. Show that the least upper bound principle implies that every Cauchy sequence converges, and conversely.

K. If $I_n = [a_n, b_n]$ is a nonempty closed interval for each $n \in \mathbb{N}$ such that $I_1 \supset I_2 \supset I_3 \supset \cdots$, show that $\bigcap_{n=1}^{\infty} I_n \neq \varnothing$.

L. If E is a bounded set of real numbers and F is a nonempty subset of E, then show that

$$\inf E \leq \inf F \leq \sup F \leq \sup E.$$

M. Show that the set of all irrational numbers is uncountable.

§5. Topological Concepts in $\mathbb{R}$

Many of the deepest properties of real analysis depend on certain topological notions and results. In this section we introduce some basic topological concepts and results which will be frequently used in this book.

Of basic importance for our later study is the *open interval*. We define the open interval (a, b) to be the set $\{x \in \mathbb{R}: a < x < b\}$. We always take $a < b$, but we consider also the infinite open intervals $(a, \infty) = \{x \in \mathbb{R}: a < x\}$ and $(-\infty, b) = \{x \in \mathbb{R}: x < b\}$. Sometimes we write $(-\infty, \infty)$ for $\mathbb{R}$. A generalization of the notion of an open interval is given by that of an open set.

5.1. Definition. A set G in ℝ is called *open* if, for each x in G, there is a positive real number r such that every y with $|x - y| < r$ belongs to G.

We can rephrase this definition by saying that a set G is open if every point in G is the center of some open interval entirely contained in G. The open intervals are examples of open sets. The real line ℝ itself is open. The empty set $\varnothing$ is open.

We now state the basic properties of open sets in ℝ.

5.2. Proposition.

(a) *The intersection of any two open sets is open.*

(b) *The union of any collection of open sets is open.*

Proof. (a) Let G_1 and G_2 be open and let $x \in G_1 \cap G_2$. Since $x \in G_1$, there exists $r_1 > 0$ such that all y with $|x - y| < r_1$ belong to G_1. Similarly, there exists $r_2 > 0$ such that all y wtih $|x - y| < r_2$ belong to G_2. Take r to be the smaller of r_1 and r_2. Then $r > 0$, and if $|x - y| < r$, then y belongs to both G_1 and G_2, i.e., to $G_1 \cap G_2$.

(b) Let $\mathscr{C}$ be a collection of open sets, and let U be their union. To show that U is open, let $x \in U$. By the definition of the union, there is an open set G in $\mathscr{C}$ such that $x \in G$. Since G is open, there is an $r > 0$ such that all y with $|x - y| < r$ belong to G, and hence to U, since $G \subset U$. Therefore, U is open. $\square$

By induction, it follows from property (a) above that the intersection of any finite collection of open sets is open. However, the intersection of an infinite collection of open sets may not be open. For example,

$$\bigcap_{n=1}^{\infty} \left(-\frac{1}{n}, \frac{1}{n} \right) = \{0\}$$

and $\{0\}$ is not an open set.

The nature of open sets in ℝ is given explicitly by the following characterization, which is due to Georg Cantor (1882):

5.3. Theorem. *Every open set of real numbers is the union of a countable collection of mutually disjoint open intervals.*

Proof. Let G be an open set and $x \in G$. Then there is a $y < x$ such that the open interval $(y, x) \subset G$. Let $a = \inf\{y : (y, x) \subset G\}$. Consider likewise $b = \sup\{z : (x, z) \subset G\}$. Evidently $a < b$. Note that a, b can be $-\infty$, ∞, respectively, but we cannot have both $a = -\infty$ and $b = \infty$ if $G \neq ℝ$. Let $I(x) = (a, b)$. Then $I(x)$ is an open interval containing x, and $I(x) \subset G$. Furthermore, we have $b \notin G$. In fact, if $b \in G$, then for some $r > 0$ we have $(b - r, b + r) \subset G$, contradicting the definition of b. Similarly, $a \notin G$.

It is easy to see that if x and y are two distinct points of G, we have either $I(x) = I(y)$ or $I(x) \cap I(y) = \varnothing$. Now consider the collection of open intervals $I(x)$, $x \in G$. Since each x in G belongs to $I(x)$, G must be the union of mutually disjoint open intervals $I(x)$. Since each $I(x)$ should contain a rational point, it follows that the number of distinct $I(x)$'s is countable (see Exercise 5D). ☐

We define the *closed interval* $[a, b]$ to be the set $\{x: a \le x \le b\}$. For closed intervals we take a and b to be finite. The generalization of the notion of a closed interval is given by that of a closed set.

5.4. Definition. A set F is called *closed* if it is the complement of an open set.

It follows that the complement of a closed set is open. From Proposition 5.2, by using De Morgan's laws, we have the following properties of closed sets:

5.5. Proposition.

(a) *The union of any two closed sets is closed.*
(b) *The intersection of any collection of closed sets is closed.*

Though a set may be simultaneously open and closed, a set may also be neither open nor closed. Both $\varnothing$ and $\mathbb{R}$ are open and closed. $[0, 1)$ is neither open nor closed.

We now introduce an additional topological notion which will permit us to characterize closed sets.

5.6. Definition. A point x is called a *cluster point* (or an *accumulation point*) of a set A if, for every $r > 0$, there is a y in A, $y \ne x$, such that $|x - y| < r$.

This is equivalent to saying that x *is a cluster point of A if every open interval containing x also contains a point of A different from x.*

Every point of the closed interval $[a, b]$ is a cluster point. Note that nothing is specified about whether or not a cluster point of a set is in the set. For example, a is a cluster point of (a, b) but $a \notin (a, b)$.

5.7. Proposition. *A set F in $\mathbb{R}$ is closed if and only if it contains every cluster point of F.*

Proof. Suppose that F is closed and $x \notin F$. We infer that x cannot be a cluster point of F because CF is open and $F \cap CF = \varnothing$. This shows that every cluster point of F is in F.

Conversely, suppose that F contains all its cluster points. We want to show that CF is open. To do this, let $y \in CF$; according to our hypothesis, y is not a cluster point of F, so there must be an open interval I containing y such that $I \cap F = \varnothing$, i.e., $I \subset CF$. This, however, means that CF is open. ☐

We say that a collection $\mathscr{C}$ of sets *covers* a set A if $A \subset \bigcup \{G: G \in \mathscr{C}\}$. The collection $\mathscr{C}$ is then called a *cover* of A. If $\mathscr{C}$ contains only open sets, we call $\mathscr{C}$ an *open cover*. If $\mathscr{C}$ contains only a finite number of sets, we call $\mathscr{C}$ a *finite cover*. If $\mathscr{C}$ is a cover of A, then a subcollection $\mathscr{C}^*$ of $\mathscr{C}$ is called a *subcover* of A if $\mathscr{C}^*$ is also a cover of A.

5.8. Definition. A set K is called *compact* if every open cover of K admits a finite subcover.

In order to apply this definition to prove that a certain set K is compact, we need to examine all possible collections of open sets whose union contains K and show that K is contained in the union of some finite subcollecton of each of these collections. It is usually not an easy task to prove that a set is compact using the definition alone.

5.9. Proposition. *The closed interval* $[a, b]$ *is compact.*

Proof. Let $\mathscr{C}$ be an open cover of $[a, b]$. Let E be the set of numbers $x \leq b$ such that the interval $[a, x]$ is contained in the union of a finite number of sets in $\mathscr{C}$. Then $E \neq \varnothing$ since $a \in E$, and E is bounded above by b. Let $c = \sup E$. Since $c \in [a, b]$, there is an open set G in $\mathscr{C}$ such that $c \in G$. Hence, for some $\varepsilon > 0$, the interval $(c - \varepsilon, c + \varepsilon) \subset G$. Now $c - \varepsilon$ is not an upper bound of E, and hence, there exists $x \in E$ with $x > c - \varepsilon$. Since $x \in E$, $[a, x]$ is contained in the union of a finite number of sets in $\mathscr{C}$. Hence, by adding the single set G to the finite number already required to cover $[a, x]$, we conclude that $c \in E$ and $d \in E$ for any d satisfying $c < d < c + \varepsilon$ and $d \leq b$. This gives a contradiction unless $c = b$. Hence $c = b$ and $b \in E$. ☐

We now present a remarkable theorem which characterizes every compact subset of ℝ. The following theorem is known variously as the Heine–Borel theorem, the Borel–Lebesgue theorem, and the Borel covering theorem, after Eduard Heine (1821–1881) (a student of Weierstrass), Emile Borel (1871–1956), and Henri Lebesgue (1875–1943).

5.10. Heine–Borel Theorem. *A subset of* ℝ *is compact if and only if it is closed and bounded.*

Proof. Suppose that K is a compact set. The proof that K is bounded is very simple. Since $K \subset \mathbb{R} = \bigcup_{m=1}^{\infty} (-m, m)$ and K is compact, there exists a natural number N such that $K \subset (-N, N)$. This proves that K is bounded.

To show that the compact set K is closed, we will prove that CK is open. Let $x \in CK$. For each $y \in K$ we can find disjoint open intervals U_y and V_y containing y and x, respectively. Then the collection $\mathscr{C}$ of all U_y, $y \in K$, becomes an open cover of K. Therefore, K admits a finite subcover. For convenience of notation let the finite subcover be $U_1, \ldots, U_n$, with U_m being the U_y

associated with a certain y_m. Let the corresponding V_y's be $V_1, \ldots, V_n$. Now let $U = \bigcup_{m=1}^{n} U_m$ and $V = \bigcap_{m=1}^{n} V_m$. Then $U \cap V = \emptyset$, $K \subset U$, and $V \subset CK$. But V is an open set containing x. Therefore, CK is open.

Conversely, let K be bounded and closed. Since K is bounded, we can enclose K in a closed interval $[a, b]$. Let $\mathscr{C}$ be an open cover of K. Then $K \subset [a, b] \subset \mathbb{R} = CK \cup K = CK \cup \bigcup \{G : G \in \mathscr{C}\}$. Therefore the collection $\mathscr{C} \cup \{CK\}$ is an open cover of $[a, b]$ since CK is open. By Proposition 5.9, $[a, b]$ is compact; thus there is a finite subcover $\mathscr{C}^*$ of $\mathscr{C} \cup \{CK\}$ which covers $[a, b]$ and hence K. Since $K \cap CK = \emptyset$, $\mathscr{C}^* \backslash \{CK\}$ covers K. However, $\mathscr{C}^* \backslash \{CK\}$ is a finite subcollection of $\mathscr{C}$. Therefore, K is compact. $\square$

The idea of the previous theorem was found in Heine's work (1872) in proving that a continuous function on $[a, b]$ is uniformly continuous (see Proposition 6.5). In 1894 Emile Borel established the theorem that a countable cover of a bounded closed set can be reduced to a finite cover in his thesis presented to the Faculté des Sciences in Paris. Henri Lebesgue (1905) extended this result to the uncountable cover of the compactness theorem in *Sur les fonctions représentables analytiquement* (in particular, see p. 176). This same extension was given simultaneously by F. Riesz (1905).

The next theorem is considered the fundamental theorem about accumulation points.

5.11. Bolzano–Weierstrass Theorem. *Every bounded infinite set has a cluster point.*

Proof. Let B be a bounded set with an infinite number of elements. Suppose that B has no cluster points. Let I be a closed interval containing B. For each $x \in I$, let $I(x)$ be an open interval containing x and only a finite number of points of B. Such an $I(x)$ can be found since x is not a cluster point of B. Then $\{I(x) : x \in I\}$ is an open cover for I. Since I is compact, I can be covered by a finite number of such intervals; but then I contains only finitely many points of B, and hence B is finite, since $B \subset I$. This contradiction shows that B must have a cluster point. $\square$

Bernard Bolzano (1781–1848), Austrian theologian, logician, and mathematician, made early and important contributions to real analysis. Bolzano's work in 1817 showed some of the ideas that underlie the preceding theorem. The present form of the theorem was first proved by Karl Weierstrass (1815–1897), a great German mathematician, in his unpublished lectures at Berlin around 1860. The tendency to insist upon complete rigor in mathematical proofs is a result, in part, of Weierstrass's influence.

The proof given above illustrates the use of compactness. A direct proof follows from the least upper bound principle (see Exercise 5M). Therefore, the Bolzano–Weierstrass theorem and the Heine–Borel theorem are equivalent.

EXERCISES 5

A. Show that $\varnothing$ is open.

B. A point x is called an *interior point* of a set G if there is an $r > 0$ such that the interval $(x - r, x + r) \subset G$. The set of interior points of G is denoted by G^0. Show that G is open if and only if $G = G^0$.

C. Prove in detail the assertion, in the proof of Proposition 5.3, that if $x \neq y$ and $I(x) \cap I(y) \neq \varnothing$, then $I(x) = I(y)$.

D. Let $\mathscr{C}$ be a collection of mutually disjoint open sets. Show that $\mathscr{C}$ is a countable collection.

E. Find an example to show that the union of a countably infinite number of closed sets is not necessarily closed.

F. Is the set $\mathbf{Q}$ of rational numbers in $\mathbf{R}$ closed, open, or neither?

G. **Definition.** Let $A \subset \mathbf{R}$. A subset B of A is called *dense in* A if every point of A is a cluster point of B.
   Show that $\mathbf{Q}$ is dense in $\mathbf{R}$.

H. Show that a finite subset of $\mathbf{R}$ has no cluster point.

I. Show that, if x is a cluster point of a set A, there exists a sequence (x_n) in A converging to x, where the x_n's are distinct.

J. Ernst Lindelöf (1870–1946). Let $\mathscr{C}$ be a collection of open sets of real numbers. Then there is a countable subcollection (G_n) of $\mathscr{C}$ such that

$$\bigcup \{G : G \in \mathscr{C}\} = \bigcup_{n=1}^{\infty} G_n.$$

K. Show that, if F is a closed set contained in a compact set, then F is also compact.

L. Let F be a compact set. Show that $\sup F \in F$ and $\inf F \in F$.

M. Give a direct proof of the Bolzano–Weierstrass theorem (5.11).

N. Let A be a bounded set. Are $\inf A$ and $\sup A$ cluster points of A?

O. Let ξ be an irrational number. Show that the set

$$\{m + n\xi : m, n \in \mathbf{Z}\}$$

   is dense in $\mathbf{R}$.

P. If $A \subset \mathbf{R}$, let $\bar{A}$ denote the intersection of all closed sets containing A. The set $\bar{A}$ is called the *closure* of A. Prove that $A = \bar{A}$ if and only if A is closed.

§6. Continuous Functions

We have already discussed the concept of functions in §2. In this section we shall be exclusively interested in those functions which have their domain and range in the real number system. Let E be a set of real numbers.

6.1. Definition. A function f is said to be *continuous at the point* x in E if, given $\varepsilon > 0$, there is a $\delta > 0$ such that for all y in E with $|x - y| < \delta$ we have $|f(x) - f(y)| < \varepsilon$. If f is continuous at every point of its domain, we say that f is a *continuous function*.

The continuity at the point x in E can also be stated as follows:

6.2. Proposition. *A function f is continuous at the point x in E if and only if, for every sequence (x_n) in E such that $x_n \to x$, we have $f(x_n) \to f(x)$.*

Proof. If f is continuous at x, then for every $\varepsilon > 0$ there is a $\delta > 0$ such that for all y in E with $|x - y| < \delta$ we have $|f(x) - f(y)| < \varepsilon$. Let (x_n) be a sequence in E such that $x_n \to x$. Then there exists a natural number N such that $n > N$ implies $|x_n - x| < \delta$. Hence $n > N$ implies $|f(x_n) - f(x)| < \varepsilon$.

Conversely, suppose that for every sequence (x_n) in E, $x_n \to x$ implies $f(x_n) \to f(x)$. Let us assume that f is not continuous at x. Then there is an $\varepsilon_0 > 0$ such that for every $\delta > 0$ there exists y in E such that $|x - y| < \delta$ and $|f(x) - f(y)| \geq \varepsilon_0$ (why?). For each $n \in \mathbb{N}$, let $\delta_n = 1/n$ and $U_n = \{y \in E: |x - y| < \delta_n$ and $|f(x) - f(y)| \geq \varepsilon_0\}$. Since each U_n is not empty, we pick $x_n \in U_n$ for each n. Then it is clear that $x_n \to x$. But $|f(x_n) - f(x)| \geq \varepsilon_0$ for all $n \in \mathbb{N}$. This contradicts the fact that $f(x_n) \to f(x)$. ☐

Now we prove some of the deeper properties of continuous functions. In particular, we have the following propositions:

6.3. Proposition. *A continuous function with a compact domain has a compact range; i.e., the continuous image of a compact set is compact.*

Proof. Let K be a compact set and let f be a continuous function on K. We show that the image $f(K)$ is bounded and closed.

First we prove that $f(K)$ is bounded. Since f is continuous on K, for any $x \in K$ and $\varepsilon = 1$ there corresponds an open interval $I(x)$ centered at x such that

$$|f(x) - f(y)| < 1$$

wherever $y \in I(x) \cap K$. Then the collection $\{I(x): x \in K\}$ is an open cover for K. Therefore, there exist a finite number of points $x_1, \ldots, x_n$ in K such that

$$K \subset I(x_1) \cup \cdots \cup I(x_n).$$

Let M be the largest among $|f(x_1)|, \ldots, |f(x_n)|$. Then for every $x \in K$, we have $x \in I(x_m)$ for some m, $1 \leq m \leq n$, and hence

$$|f(x)| < |f(x_m)| + 1 \leq M + 1.$$

This proves that $f(K)$ is bounded.

Next we prove that $f(K)$ is closed. If $f(K)$ is a finite set, then it is clearly closed. Assume that $f(K)$ is an infinite set. Let w be a cluster point of $f(K)$.

Then there exists a sequence (x_n) in K such that $f(x_n) \to w$ and all $f(x_n)$ are distinct (see Exercise 5I). Therefore the set $\{x_n: n \in \mathbb{N}\}$ should have a cluster point x by the Bolzano–Weierstrass theorem. Since K is closed, x must be in K by Proposition 5.7. By Exercise 5I again, the sequence (x_n) contains a subsequence (x_{n_k}) converging to x. Since f is continuous at x, we have $f(x_{n_k}) \to f(x)$ by Proposition 6.2. On the other hand, $(f(x_{n_k}))$ is a subsequence of $(f(x_n))$ and hence itself converges to w, $f(x_{n_k}) \to w$. Therefore $w = f(x)$; that is, $w \in f(K)$, which, by Proposition 5.7, proves that $f(K)$ is closed. □

6.4. Proposition. *Every continuous function f with a compact domain K has a maximum and a minimum; that is, there are points x_1 and x_2 in K such that $f(x_1) \le f(x) \le f(x_2)$ for all x in K.*

Proof. Since f is continuous on K, the range $f(K)$ is compact, according to the preceding proposition. Let m and M be the greatest lower bound and the least upper bound of $f(K)$, respectively. These exist since $f(K)$ is bounded. Our goal is to show that there are points x_1 and x_2 in K such that $m = f(x_1)$ and $M = f(x_2)$. By the very definition of m, any open interval containing m will contain at least one point in $f(K)$ (why?). If m is in $f(K)$, we have nothing to show. Otherwise m will be a cluster point of $f(K)$. But $f(K)$ is closed, and hence $m \in f(K)$; that is, there is a point $x_1 \in K$ such that $m = f(x_1)$. Similarly, there is an x_2 in K such that $M = f(x_2)$. □

Let us return to Definition 6.1 of a continuous function and observe that δ depends, in general, on both ε and x. That δ depends on x is based on the fact that the function f may change its values rapidly in the open interval $(y - \delta, y + \delta)$ for some points y. Now, it can happen that a continuous function behaves such that the number δ can be chosen to be independent of the point in the domain of f, that is, depending only on ε.

6.5. Definition. A function $f: E \to \mathbb{R}$ is said to be *uniformly continuous on E* if, given $\varepsilon > 0$, there is a $\delta > 0$ such that for all x, y in E with $|x - y| < \delta$, we have $|f(x) - f(y)| < \varepsilon$.

It is clear that if f is uniformly continuous on E, then it is continuous on E. In general, the converse does not hold. For example, $f(x) = 1/x$ is not uniformly continuous on $\{x: x > 0\}$.

We now present Heine's theorem about continuous functions.

6.6. Theorem (Heine, 1872). *If f is continuous on a compact set K, then f is uniformly continuous on K.*

Proof. For each $x \in K$ and $\varepsilon > 0$ there exists $\delta(\varepsilon, x) > 0$ [the notation $\delta(\varepsilon, x)$ means that the number $\delta(\varepsilon, x)$ depends on ε and x] such that if $|x - y| < \delta(\varepsilon, x)$ then $|f(y) - f(x)| < \varepsilon/2$.

Now for each $x \in K$, let $I(x) = \{y: |y - x| < \delta(\varepsilon, x)/2\}$. Then the collection

$$\mathscr{C} = \{I(x): x \in K\}$$

is an open cover for the compact set K, and hence there are finitely many points $x_1, \ldots, x_n$ in K such that

$$K \subset I(x_1) \cup \cdots \cup I(x_n).$$

Let $\delta = \min\{\delta(\varepsilon, x_k)/2: k = 1, \ldots, n\}$. Suppose that $x, y \in K$ and $|x - y| < \delta$. Then x belongs to some $I(x_k)$. This implies that

$$|y - x_k| \le |y - x| + |x - x_k| < \delta + \delta(\varepsilon, x_k)/2 \le \delta(\varepsilon, x_k).$$

Hence

$$|f(y) - f(x)| \le |f(y) - f(x_k)| + |f(x_k) - f(x)| < \varepsilon$$

which proves that f is uniformly continuous on K.   □

EXERCISES 6

A. If $f: \mathbb{R} \to \mathbb{R}$ is continuous on $\mathbb{R}$ and if $f(a) > 0$, show that there is an open interval I containing x such that f is positive on I. Does the same conclusion follow if f is only continuous at the point a?

B. Give an example of a bounded and continuous function $f: \mathbb{R} \to \mathbb{R}$ which does not have a maximum and a minimum.

C. Show that the function $D: \mathbb{R} \to \mathbb{R}$ defined by

$$D(x) = \begin{cases} 1 & \text{if } x \in \mathbb{Q}, \\ 0 & \text{if } x \notin \mathbb{Q}, \end{cases}$$

is nowhere continuous. (This function is called the *Dirichlet function*.)

D. Suppose that $f: (0, 1) \to \mathbb{R}$ is continuous. Can f be defined at $x = 0$ and $x = 1$ in such a way that it becomes continuous on $[0, 1]$?

E. Suppose that $f: (0, 1) \to \mathbb{R}$ is uniformly continuous. Can f be defined at $x = 0$ and $x = 1$ in such a way that it becomes continuous on $[0, 1]$?

F. Let E be a set with the property that every continuous function with domain E is uniformly continuous. Is E necessarily compact?

G. If $f: \mathbb{R} \to \mathbb{R}$ is such that $f(x + y) = f(x) + f(y)$ for all $x, y \in \mathbb{R}$.
   (i) Prove that $f(x) = xf(1)$ for all rational x.
   (ii) Prove that $f(x) = xf(1)$ for all $x \in \mathbb{R}$ if f is continuous at $x = 0$.

§7. Metric Spaces

A metric space is a set in which we can speak of the distance between two points. It is a generaliztion of the real line where, in making the generalization, only some of the geometric properties have been preserved. For $x, y \in \mathbb{R}$

the geometric interpretation of $|x - y|$ is the distance from x to y. If we define the distance function d by

$$d(x, y) = |x - y|,$$

then we have the following consequences for any points x, y, z in $\mathbb{R}$:

$$d(x, x) = 0,$$

$$d(x, y) > 0 \quad \text{if} \quad x \neq y,$$

$$d(x, y) = d(y, x),$$

$$d(x, y) \leq d(x, z) + d(z, y) \quad \text{(triangle inequality)}.$$

This is a motivation of the following definition:

7.1. Definition. A *metric space* is a set M with a function $d\colon M \times M \to \mathbb{R}$ such that:

(a) $d(x, x) = 0$;
(b) $d(x, y) > 0$ if $x \neq y$;
(c) $d(x, y) = d(y, x)$; and
(d) $d(x, y) \leq d(x, z) + d(z, y)$ (triangle inequality);*

where x, y, $z \in M$. The function d is called a *metric* for M.

An immediate consequence of this definition is the property:

(e) $|d(x, y) - d(x, z)| \leq d(y, z)$.

The proof is left to the reader.

Here are some examples of metric spaces.

7.2. Example. Let $\mathbb{R}^n = \mathbb{R} \times \cdots \times \mathbb{R}$ (n times) and define

$$d(x, y) = \left(\sum_{k=1}^{n} |x_k - y_k|^2 \right)^{1/2},$$

where $x = (x_1, \ldots, x_n)$, $y = (y_1, \ldots, y_n)$.

Relations (a), (b), and (c) are obvious, but the triangle inequality (d) requires discussion. We begin by proving the *Cauchy inequality*, a special case of the Cauchy–Bunyakovskii–Schwarz inequality (see §1, Chapter VI).

Cauchy Inequality.

$$\left(\sum_{k=1}^{n} x_k y_k \right)^2 \leq \left(\sum_{k=1}^{n} x_k^2 \right) \left(\sum_{k=1}^{n} y_k^2 \right).$$

* *Any ass knows this.* Simply put a haystack at one corner of a triangle and an ass at another. The ass will certainly not go along two sides of the triangle to get his hay [Euclid (circa 300 B.C.), 1956, p. 287].

Proof. Clearly, for any real number c, we have

$$\sum_{k=1}^{n} (x_k + cy_k)^2 \geq 0.$$

This is equivalent to

$$\sum_{k=1}^{n} x_k^2 + 2c \sum_{k=1}^{n} x_k y_k + c^2 \sum_{k=1}^{n} y_k^2 \geq 0.$$

If $\sum_{k=1}^{n} y_k^2 \neq 0$, we let

$$c = \frac{-\sum_{k=1}^{n} x_k y_k}{\sum_{k=1}^{n} y_k^2}$$

and the Cauchy inequality follows. If $\sum_{k=1}^{n} y_k^2 = 0$ and $\sum_{k=1}^{n} x_k^2 \neq 0$, we can interchange the roles of $(x_1, \ldots, x_n)$ and $(y_1, \ldots, y_n)$. If both are zero, the inequality reduces to $0 = 0$. $\square$

Minkowski Inequality.

$$\left[\sum_{k=1}^{n} (x_k + y_k)^2 \right]^{1/2} \leq \left(\sum_{k=1}^{n} x_k^2 \right)^{1/2} + \left(\sum_{k=1}^{n} y_k^2 \right)^{1/2}.$$

Proof.

$$\sum_{k=1}^{n} (x_k + y_k)^2 = \sum_{k=1}^{n} x_k^2 + 2 \sum_{k=1}^{n} x_k y_k + \sum_{k=1}^{n} y_k^2$$

$$\leq \sum_{k=1}^{n} x_k^2 + 2 \left(\sum_{k=1}^{n} x_k^2 \sum_{k=1}^{n} y_k^2 \right)^{1/2} + \sum_{k=1}^{n} y_k^2$$

$$\leq \left[\left(\sum_{k=1}^{n} x_k^2 \right)^{1/2} + \left(\sum_{k=1}^{n} y_k^2 \right)^{1/2} \right]^2.$$

Hence we have the Minkowski inequality. $\square$

The triangle inequality follows at once from the Minkowski inequality if we replace x_k by $x_k - z_k$ and y_k by $z_k - y_k$. Therefore $\mathbb{R}^n$ is a metric space. This space is called *n-dimensional Euclidean space*.

7.3. Example. We now introduce the space $C[a, b]$ of all continuous real-valued functions on $[a, b]$. The function d defined by

$$d(f, g) = \sup\{|f(x) - g(x)|: x \in [a, b]\}$$

is a metric. In fact, d obviously satisfies relations (a), (b), and (c), but relation (d) is also satisfied, since

$$|f(x) - h(x)| = |f(x) - g(x) + g(x) - h(x)|$$

$$\leq |f(x) - g(x)| + |g(x) - h(x)|,$$

and hence

$$\sup|f(x) - h(x)| \leq \sup|f(x) - g(x)| + \sup|g(x) - h(x)|,$$

which proves the validity of the triangle inequality.

As well as using the idea of a metric on the real line, one learns early in calculus that convergence of sequences is of vital importance for the development of analysis. We shall model our definition of the convergence of sequences in a general metric space after the convergence of real sequences.

7.4. Definition. Let M be a metric space with metric d. A sequence (x_n) in M is called *convergent* if there exists $x \in M$ such that $d(x_n, x) \to 0$ as $n \to \infty$; i.e., for any $\varepsilon > 0$, there exists $N > 0$ such that $m \geq N$ implies $d(x_m, x) < \varepsilon$. We then write $x = \lim_{n \to \infty} x_n$, $x_n \to x$, or $x = \lim x_n$ and call x the *limit* of the sequence (x_n).

Our first proposition tells us that a sequence cannot have more than one limit.

7.5. Proposition. *A convergent sequence in a metric space has a unique limit.*

Proof. If $x_n \to x$ and also $x_n \to y$ in the metric d, then by the triangle inequality

$$0 \leq d(x, y) \leq d(x, x_n) + d(x_n, y) \to 0$$

as $n \to \infty$. Hence $d(x, y) = 0$, so $x = y$; i.e., the limit is unique. $\square$

In §4 we defined a Cauchy sequence. We shall now make the obvious definition of a Cauchy sequence in a metric space.

7.6. Definition. A sequence (x_n) in a metric space M is called a *Cauchy sequence* if, given $\varepsilon > 0$, there is an N such that for all $n > N$ and all $m > N$ we have $d(x_m, x_n) < \varepsilon$.

7.7. Proposition.

(a) *Every convergent sequence is a Cauchy sequence.*
(b) *If a Cauchy sequence has a convergent subsequence, then the whole sequence is conergent.*

Proof. (a) Let (x_n) be convergent, say $x_n \to x$. Then if $\varepsilon > 0$ there is an N such that $d(x, x_n) < \varepsilon/2$ for all $n > N$. Since

$$d(x_m, x_n) \leq d(x, x_n) + d(x, x_m)$$

it follows that $d(x_m, x_n) < \varepsilon$ if $m > N$ and $n > N$, so that (x_n) is Cauchy.

(b) Let (x_n) be a Cauchy sequence and suppose that (x_{n_k}) is a convergent subsequence. By this we mean

$$(x_{n_k}) = (x_{n_1}, x_{n_2}, \ldots),$$

where $n_1 < n_2 < \cdots$ are natural numbers and also $x_{n_k} \to x$ as $k \to \infty$ in the metric d. Thus we have

$$0 \leq d(x_n, x) \leq d(x_n, x_{n_k}) + d(x_{n_k}, x). \qquad (*)$$

On the other hand, for any given $\varepsilon > 0$, there is an N such that for $n_k > N$ and $n > N$ we have

$$d(x_{n_k}, x) < \frac{\varepsilon}{2} \quad \text{and} \quad d(x_n, x_{n_k}) < \frac{\varepsilon}{2}.$$

Therefore, from $(*)$ we have $d(x_n, x) < \varepsilon$ for all $n > N$. This proves that the Cauchy sequence (x_n) converges to the limit of the convergent subsequence (x_{n_k}). $\qquad \square$

We shall find part (b) of Proposition 7.7 useful in Chapter IV. Although every convergent sequence is a Cauchy sequence, *it is not in general true that a Cauchy sequence converges.* For example, if $M = (0, 1)$ and d is the usual metric for $\mathbb{R}$, i.e., $d(x, y) = |x - y|$, then in the metric space M the sequence $(1/n)$ is a Cauchy sequence with respect to the metric d but does not converge. This sequence (x_n) fails to converge because, roughly speaking, the point that it ought to converge to (0) is missing from the set M. Another simple example is the space $\mathbb{Q}$ with the usual metric $d(x, y) = |x - y|$. The sequence

$$1, 1.4, 1.41, 1.414, 1.4142, \ldots$$

(the truncations of the infinite decimal representing $\sqrt{2}$) is a Cauchy sequence not converging to any rational number. These facts lead us to the following definition.

7.8. Definition. If a metric space M has the property that every Cauchy sequence converges to some point of the space, we say that the space is *complete.*

An obvious example of a complete metric space is $\mathbb{R}$. In general, the n-dimensional Euclidean space $\mathbb{R}^n$ is complete (see Exercise 7C).

The space $C[a, b]$ in Example 7.3 is complete. The proof is left to the reader (see Exercise 7D). Other examples of complete metric spaces are given by the Banach spaces discussed in Chapter VI.

EXERCISES 7

A. Prove that

$$|d(x, y) - d(x, z)| \leq d(y, z)$$

for any three points x, y, and z in a metric space M.

B. Let M be a set. Define $d: M \times M \to \mathbb{R}^+$ by

$$d(x, y) = \begin{cases} 1 & \text{if } x \neq y, \\ 0 & \text{if } x = y. \end{cases}$$

Show that d is a metric. (The space M with metric d is called *discrete*.)

C. Show that the n-dimensional Euclidean space $\mathbb{R}^n$ is complete.

D. Let (f_n) be a Cauchy sequence in $C[a, b]$.
   (a) Show that for each $x \in [a, b]$, the sequence $(f_n(x))$ of real numbers converges, say, to $f(x)$.
   (b) Show that $f: [a, b] \to \mathbb{R}$ is continuous on $[a, b]$.

E. Let $P[0, 1]$ be the set of all real polynomials with metric

$$d(f, g) = \sup\{|f(x) - g(x)|: x \in [0, 1]\}$$

Show that the metric space $P[0, 1]$ is not complete. (*Hint*: Consider $\exp(x) = 1 + x + x^2/2 + \cdots$.)

CHAPTER I

The Riemann Integral

In this chapter we study elementary integration theory for functions defined on closed intervals. Although we expect that the reader has had experience with integral calculus and that the ideas are familiar, we shall not require any special results to be known. For pedagogical reasons we shall first treat the Cauchy integral. After this has been done, we will study in §3 the Riemann integral. Our attention here is focused exclusively on the definition and existence, since these concepts are often mysterious even to students who have ample knowledge of the numerous applications and techniques of evaluating Riemann integrals from their study of calculus.

The purpose of this chapter is to present motivation for the Lebesgue inegral through a historical development of the concepts of integration.

§1. The Cauchy Integral

Before Augustin-Louis Cauchy (1789–1857), one merely defined integrals geometrically, showing which areas had to be added or subtracted in order to obtain the integral $\int_a^b f(x)\,dx$. For Cauchy a definition was necessary because of his concern for the logical foundations of mathematical analysis. His predecessors, in general, justified mathematical analysis by means of its physical interpretation. Cauchy wrote an important book, the *Cours d'Analyse de l'École Royal Polytechnique*, in 1821. This work was based on his lectures in analysis, which he had given at the École Polytechnique and the other Paris colleges at which he taught. In *Cours d'Analyse*, he set forth a new concept of continuity which has remained standard ever since. In this *Résumé des Leçons Données a l'École Royal Polytechnique sur le Calcul Infinitésimale* (1823), Cauchy defined the definite integral of a continuous function over a closed interval $[a, b]$ in about the same way we do today.

Before giving the definition, we must introduce some auxiliary notions. By a *partition P* of the closed interval $[a, b]$ we mean a finite ordered set

$$P = \{a = x_0 < x_1 < \cdots < x_n = b\}.$$

The *norm* $|P|$ of the partition is

$$|P| = \sup\{x_j - x_{j-1}: 1 \leq j \leq n\}.$$

A *refinement P'* of P is a partition of $[a, b]$ such that $P' \supset P$. It is clear that

$$|P'| \leq |P|.$$

Let f be a continuous function on $[a, b]$, and consider a partition P of the interval,

$$P: \quad a = x_0 < x_1 < \cdots < x_n = b.$$

The *Cauchy sum* is defined by

$$S(P; f) = \sum_{j=1}^{n} f(\xi_j)(x_j - x_{j-1}),$$

where $x_{j-1} \leq \xi_j \leq x_j$ (see Figure 1.1).

Note that the ξ_j are arbitrary in that ξ_j can be any point whatsoever of $[x_{j-1}, x_j]$. The expression $S(P; f)$ is slightly inadequate in that it does not show the dependence of $S(P; f)$ on the set $\{\xi_1, \ldots, \xi_n\}$. However, the gain in precision by indicating this dependence is not worth the increased notational cumbersomeness. To emphasize the fact that $S(P; f)$ is not uniquely determined, we use the phrase *any possible Cauchy sum* $S(P; f)$ *relative to P*.

The definite integral of the function f on $[a, b]$ is defined as the limit of the Cauchy sums $S(P; f)$ as $|P| \to 0$. We shall first explain what is meant by such

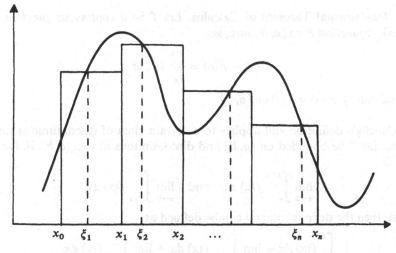

Figure 1.1

a limit. We denote this limit by $\lim_{|P|\to 0} S(P; f)$. The statement

$$\lim_{|P|\to 0} S(P; f) = L,$$

where L is a real number, means that for any $\varepsilon > 0$ there is a $\delta > 0$ such that for any partition P with $|P| < \delta$ and for any possible Cauchy sum $S(P; f)$ relative to P, the inequality

$$|S(P; f) - L| < \varepsilon \qquad\qquad (*)$$

is satisfied.

Equivalently, $(*)$ means that for any $\varepsilon > 0$ there is a partition P_ε such that

$$|S(P; f) - L| < \varepsilon$$

holds for all partitions $P \supset P_\varepsilon$ where $S(P; f)$ is any possible Cauchy sum relative to P. The proof is left to the reader (see Exercise 1A).

If $\lim_{|P|\to 0} S(P; f)$ exists, then it is simple to show that the limit is unique (see Exercise 1B). This unique limit is called the *definite integral* of f on $[a, b]$ and the limit is written

$$\int_a^b f(x)\, dx \quad \text{or} \quad \int_a^b f.$$

Then we also say that f is *integrable* on $[a, b]$.

Using various theorems of the *Cours d'Analyse*, in particular using the continuity of f or, more precisely, the uniform continuity of f, Cauchy was able to prove the following assertion.

1.1. Proposition. *If f is continuous on $[a, b]$, f is integrable on $[a, b]$.*

One of the principal advantages of Cauchy's definition was that it enabled him to prove the fundamental theorem of integral calculus.

1.2. Fundamental Theorem of Calculus. *Let f be a continuous function on $[a, b]$. A function F on $[a, b]$ satisfies*

$$F(x) - F(a) = \int_a^x f(t)\, dt$$

if and only if $F'(x) = f(x)$ on $[a, b]$.

Cauchy's definition still applies to a certain class of discontinuous functions. Let f be bounded on $[a, b]$ and discontinuous at c in (a, b). If, for all $\varepsilon > 0$,

$$\lim_{\varepsilon \to 0} \int_a^{c-\varepsilon} f(x)\, dx \quad \text{and} \quad \lim_{\varepsilon \to 0} \int_{c+\varepsilon}^b f(x)\, dx$$

exist, then the definite integral can be defined as

$$\int_a^b f(x)\, dx = \lim_{\varepsilon \to 0} \int_a^{c-\varepsilon} f(x)\, dx + \lim_{\varepsilon \to 0} \int_{c+\varepsilon}^b f(x)\, dx.$$

Although the definite integral for a function with any finite number of discontinuities in $[a, b]$ can be defined analogously, this approach is not suitable for fucntions with an infinite number of discontinuities in $[a, b]$. Cauchy's definition is mainly for functions f which are *piecewise continuous*; i.e., f has finitely many discontinuities.

By a similar limit-taking technique, Cauchy's definition can be extended to integrals over infinite intervals:

$$\int_a^\infty f = \lim_{b \to \infty} \int_a^b f.$$

EXERCISES 1

A. Show that $\lim_{|P| \to 0} S(P; f) = L$ if and only if for any $\varepsilon > 0$ there is a partition P_ε such that the inequality

$$|S(P; f) - L| < \varepsilon$$

holds for all partitions $P \supset P_\varepsilon$ and for any possible Cauchy sum $S(P; f)$ relative to P.

B. If $\lim_{|P| \to 0} S(P; f)$ exists, show that the limit is unique.

C. If f and g are continuous on $[a, b]$ and $f(x) \le g(x)$ there, show that

$$\int_a^b f \le \int_a^b g.$$

D. If f is continuous on $[a, b]$, and if g is defined on $[a, b]$ and equal to f at every point of $[a, b]$ except for at most finitely many points, show that g is integrable on $[a, b]$ and

$$\int_a^b g = \int_a^b f.$$

E. Prove Proposition 1.1.

F. Prove Proposition 1.2.

G. **First Mean Value Theorem for Integrals.** *If f is continuous on $[a, b]$, then there exists c in (a, b) such that*

$$\int_a^b f = f(c)(b - a).$$

§2. Fourier Series and Dirichlet's Conditions

We define a trigonometric series to be a series of the form

$$\frac{1}{2}a_0 + \sum_{k=1}^n (a_k \cos kx + b_k \sin kx), \tag{1}$$

where the coefficients $a_0, a_1, \ldots, b_1, b_2, \ldots$ are real numbers. The factor $\frac{1}{2}$ is added to a_0 for convenience. The study of such series, in particular, the problem of representing a given function by a trigonometric series, originated in such physical problems as oscillations and the theory of heat conduction. These studies have been carried on since 1740.

It is easy to see that if the series (1) does converge to a sum $S(x)$, say, then for any natural number n

$$S(x + 2n\pi) = S(x),$$

so that we need only study trigonometric series in an interval of length 2π; for example, we might make it $[-\pi, \pi]$ or $[0, 2\pi]$.

The natural question which now arises is whether it is possible to represent a prescribed function f on $[-\pi, \pi]$ by a trigonometric series. Suppose that there exist sequences (a_n) and (b_n) of real numbers such that the series of the form (1) does converge to a sum $f(x)$ so that we write

$$f(x) = \tfrac{1}{2}a_0 + \sum_{k=1}^{\infty} (a_k \cos kx + b_k \sin kx). \tag{2}$$

By using elementary properties of the trigonometric functions we can now readily determine a_k, b_k in terms of $f(x)$.

The following identities are elementary:

$$\int_{-\pi}^{\pi} \sin mx \sin nx \, dx = \begin{cases} 0 & \text{if } m \neq n, \\ \pi & \text{if } m = n, \end{cases}$$

$$\int_{-\pi}^{\pi} \cos mx \cos nx \, dx = \begin{cases} 0 & \text{if } m \neq n, \\ \pi & \text{if } m = n, \end{cases}$$

$$\int_{-\pi}^{\pi} \sin mx \cos nx \, dx = 0,$$

$$\int_{-\pi}^{\pi} \cos nx \, dx = 0,$$

$$\int_{-\pi}^{\pi} \sin nx \, dx = 0.$$

If therefore we multiply both sides of equation (2) by $\cos nx$ and assume that term-by-term integration is allowed, we find

$$a_n = \frac{1}{\pi} \int_{-\pi}^{\pi} f(x) \cos nx \, dx, \qquad n = 0, 1, 2, \dots. \tag{3a}$$

On the other hand, if we multiply both sides of equation (2) by $\sin nx$ and integrate, we get

$$b_n = \frac{1}{\pi} \int_{-\pi}^{\pi} f(x) \sin nx \, dx, \qquad n = 0, 1, 2, \dots. \tag{3b}$$

The coefficients $a_0, a_1, \dots; b_1, \dots$ defined by the equations (3a) and (3b) are called the *Fourier coefficients* of f.

The above calculation of the Fourier coefficients is based on the assumption that it is known that the function f is represented as the sum of a convergent trigonometric series. Suppose now that a function f is defined on $[-\pi, \pi]$, and we can find its Fourier coefficients from the equations (3a) and (3b). Then we may write *formally*

$$f(x) \sim \tfrac{1}{2}a_0 + \sum_{k=1}^{\infty} (a_k \cos kx + b_k \sin kx). \qquad (4)$$

The series on the right side of (4) is called the *Fourier series* of f. The symbol $\sim$ is used to indicate that $f(x)$ is not necessarily equal to the series on the right. Indeed, the series on the right may diverge, or if it converges, it may converge to some function other than f.

In 1811, Joseph Fourier (1768–1830) announced his belief in the possibility of trigonometric series representation for a function. His *La Théorie Analytique de la Chaleur* was finally published in 1822. Fourier claimed then that any bounded function f defined on $[-\pi, \pi]$ can be represented by its Fourier series. He did not give a rigorous analytic proof showing that the Fourier series of f converges to f, but instead he justified the mathematics by means of its physical interpretation. His claim is not always true; however, the problem which naturally suggest itself is, under what conditions does the Fourier series of f converge to f.

The German mathematician J.P.G. Lejeune-Dirichlet (1805–1859) (a student of Fourier) initiated rigorous investigation into the theory of Fourier series. He approached the problem by considering the behavior of the partial sums

$$S_n(x) = \tfrac{1}{2}a_0 + \sum_{k=1}^{n} (a_k \cos kx + b_k \sin kx), \qquad (5)$$

where $a_0, a_1, \ldots; b_1, \ldots$ are the Fourier coefficients of f.

2.1. Proposition (Dirichlet, 1829).

$$S_n(x) = \frac{1}{2\pi} \int_{-\pi}^{\pi} f(t) \frac{\sin(n + \tfrac{1}{2})(t - x)}{\sin \tfrac{1}{2}(t - x)} \, dt.$$

The integral above is called the *Dirichlet integral*.

Proof. Substituting equations (3a) and (3b) into (5) and interchanging the order of the summation and the integration, we have

$$\pi S_n(x) = \int_{-\pi}^{\pi} f(t) \left[\frac{1}{2} + \sum_{k=1}^{n} (\cos kx \cos kt + \sin kx \sin kt) \right] dt$$

$$= \int_{-\pi}^{\pi} f(t) \left[\frac{1}{2} + \sum_{k=1}^{n} \cos k(t - x) \right] dt.$$

We now use the identity (Exercise 2B)

$$\frac{1}{2} + \sum_{k=1}^{n} \cos 2k\alpha = \frac{\sin(2n+1)\alpha}{2\sin\alpha}.$$

After substituting $2\alpha = t - x$, we have the desired identity.   □

In the preceding proposition we assumed that the function f satisfies a condition that would guarantee the existence of the definite inegrals of f and its products with $\cos kx$ and $\sin kx$ so that the Fourier coefficients a_k and b_k would possess a meaning. In his "Sur la convergence des séries trigonométriques qui servent a représenter une fonction arbitraire entre des limits données" (1829), Dirichlet explicitly assumed that the bounded function f satisfies the following two conditions:

(a) *f has only a finite number of discontinuities in $[-\pi, \pi]$.*

(b) *f has only a finite number of maxima and minima in $[-\pi, \pi]$.*

The conditions (a) and (b) for f are called *Dirichlet's conditions*. We see that they are quite restrictive, but most of the functions which arise in mathematical physics satisfy the conditions. Dirichlet then established the first rigorous proof of the following thorem concerning convergence of the Fourier series of f:

2.2. Theorem (Dirichlet, 1829). *If f is defined and bounded in $[-\pi, \pi]$ and if f satisfies Dirichlet's conditions, then*

$$S_n(x) \to \begin{cases} \frac{1}{2}[f(x^+) + f(x^-)] & \text{for } x \in (-\pi, \pi), \\ \frac{1}{2}[f(-\pi^+) + f(\pi^-)] & \text{for } x = \pm\pi, \end{cases}$$

where

$$f(x^+) = \lim_{h \to 0} f(x + h), \qquad h > 0,$$

$$f(x^-) = \lim_{h \to 0} f(x - h), \qquad h > 0.$$

An examination of his proof reveals that if the integral concept could be extended to more general functions f, then Dirichlet's proof would still work for f. This study of the convergence problem of Fourier series caused a thorough revision and reformulation of the whole theory of integration.

We close this section by giving an example of a function which does not satisfy Dirichlet's conditions. This example is from Dirichlet's paper (1829).

2.3. Example. Let c, d be distinct real numbers. Define a function D by

$$D(x) = \begin{cases} c & \text{if } x \in \mathbb{Q}, \\ d & \text{if } x \notin \mathbb{Q}, \end{cases}$$

and call it *Dirichlet's function*. Then D is nowhere continuous.

It seems this was the first example of a function that is discontinuous on an infinite set of points in a finite interval, and Dirichlet was the first mathematician to call attention to the problem of extending the concept of the integral to functions of this nature.

EXERCISES 2

A. Find the Fourier series of the following functions, each defined by the prescribed formula over $[-\pi, \pi)$ and defined elsewhere so as to have period 2π:
   (1) $f(x) = x$.
   (2) $f(x) = |x|$.
   (3) $f(x) = \begin{cases} -1 & \text{if } -\pi \leq x < 0, \\ 0 & \text{if } x = 0, \\ 1 & \text{if } 0 < x < \pi. \end{cases}$

B. Establish the identity
   $$\frac{1}{2} + \sum_{k=1}^{n} \cos 2k\alpha = \frac{\sin(2n+1)\alpha}{2\sin\alpha}.$$

C. (a) Find the Fourier series of the function
   $$f(x) = \begin{cases} x^2 & \text{for } -\pi \leq x < 0, \\ \pi^2 & \text{for } 0 \leq x < \pi, \end{cases}$$

   with $f(x + 2\pi) = f(x)$.
   (b) Deduce from (a) that
   $$\sum_{n=1}^{\infty} \frac{1}{n^2} = \frac{\pi^2}{6}.$$

D. Show that
   $$f(x) = \sin\frac{1}{x}$$

   does not satisfy Dirichlet's conditions.

E. Show that Dirichlet's function is nowhere continuous.

§3. The Riemann Integral

In 1854, Bernard Riemann (1826–1866), a German, submitted three inaugural dissertations to the University of Göttingen for a university lectureship. The three papers were in the areas of the theory of a complex variable, non-Euclidean geometry, and trigonometric series.

His main dissertation was about trigonometric series under the title, "Über die Darstellbarkeit einer Funktion durch eine trigonometrische Reihe" (On the developability of a function by a trigonometric series) (Riemann, 1866). In this paper Riemann extends Cauchy's definition of the integral by recognizing the nonessential nature of the requirement that the integrand be continuous. He replaced the continuity requirement with the weaker one that the Cauchy sums all converge to a unique limit.

Let f be a bounded function on $[a, b]$. For a partition P of $[a, b]$, we write

$$S(P; f) = \sum_{j=1}^{n} f(\xi_j)(x_j - x_{j-1}) \tag{1}$$

a Cauchy sum as in §1. Recall that $S(P; f)$ is not uniquely determined by P and f as remarked in §1. If $\lim_{|P| \to 0} S(P; f)$ exists as a unique number, then f is said to be *integrable on* $[a, b]$ *in the Riemann sense*, or simply *Riemann integrable* on $[a, b]$. The limit $\lim_{|P| \to 0} S(P; f)$ is called the *definite (Riemann) integral* of f on $[a, b]$ and is denoted again by

$$\int_a^b f(x)\, dx \quad \text{or} \quad \int_a^b f.$$

For later use we present here a necessary and sufficient condition for the Riemann integrability of a function as shown by Riemann himself in 1854. The original form of the condition is given in Exercise 3C. We shall present here his condition in the equivalent and more standard form of modern texts as described by the Frenchman Gaston Darboux (1842–1917) in his work "Mémoire sur la théorie des fonctions discontinues" (1875).

For a bounded function f on $[a, b]$ we associate with every partition P of $[a, b]$, $P: a = x_0 < x_1 < \cdots < x_n = b$, the *upper* and *lower Darboux sums*:

$$\bar{S}(P; f) = \sum_{j=1}^{n} M_j(x_j - x_{j-1}), \qquad M_j = \sup\{f(x): x_{j-1} \le x \le x_j\}, \tag{2}$$

$$\underline{S}(P; f) = \sum_{j=1}^{n} m_j(x_j - x_{j-1}), \qquad m_j = \inf\{f(x): x_{j-1} \le x \le x_j\}. \tag{3}$$

If the partition P is replaced by a finer partition P', i.e., $P' \supset P$, then

$$\underline{S}(P; f) \le \underline{S}(P'; f) \le \bar{S}(P'; f) \le \bar{S}(P; f). \tag{4}$$

We define

$$\overline{\int_a^b} f(x)\, dx = \inf \bar{S}(P; f), \tag{5}$$

$$\underline{\int_a^b} f(x)\, dx = \sup \underline{S}(P; f), \tag{6}$$

the inf and sup being taken over all partitions. The left members of (5) and (6) are called the *upper* and *lower Darboux integrals* of f over $[a, b]$, respectively. The lower and upper Darboux integrals always exist for a bounded function by (4).

If $\xi_j \in [x_{j-1}, x_j], j = 1, 2, \ldots, n$, then

$$\underline{S}(P; f) \le \sum f(\xi_j)(x_j - x_{j-1}) \le \bar{S}(P; f).$$

Therefore,

$$\underline{S}(P; f) < S(P; f) \le \bar{S}(P; f) \tag{7}$$

for any possible Cauchy sum $S(P; f)$ relative to P.

If f is Riemann integrable, then it is clear that

$$\int_{\underline{a}}^{b} f(x)\,dx \le \int_{a}^{b} f(x)\,dx \le \int_{a}^{\bar{b}} f(x)\,dx.$$

3.1. Theorem (Darboux, 1875). *The Riemann integral $\int_{a}^{b} f(x)\,dx$ exists if and only if $\int_{\underline{a}}^{b} f(x)\,dx = \int_{a}^{\bar{b}} f(x)\,dx$.*

Before proving the Darboux theorem, we recall that $\lim_{|P| \to 0} S(P; f)$ exists as a unique number $\int_{a}^{b} f(x)\,dx$ if and only if for every $\varepsilon > 0$ there exists a partition P_{ε} of $[a, b]$ such that $|S(P; f) - \int_{a}^{b} f(x)\,dx| < \varepsilon$ for any partition $P \supset P_{\varepsilon}$ and for any possible $S(P; f)$ relative to P.

Proof. Suppose that the Riemann integral of f exists. Then f must be bounded (see Exercise 3B), and for any $\varepsilon > 0$ there is a partition $P: a = x_0 < x_1 < \cdots < x_n = b$ such that

$$-\frac{\varepsilon}{2} < S(P; f) - \int_{a}^{b} f(x)\,dx < \frac{\varepsilon}{2} \tag{8}$$

for any possible Cauchy sum $S(P; f)$ relative to P. Choose ζ_j, $x_{j-1} < \zeta_j < x_j$ so that

$$M_j - f(\zeta_j) < \frac{\varepsilon}{2(b - a)}.$$

Then

$$\bar{S}(P; f) - \sum_{j=1}^{n} f(\zeta_j)(x_j - x_{j-1}) = \sum_{k=1}^{n} [M_j - f(\zeta_j)](x_j - x_{j-1}) < \frac{\varepsilon}{2}. \tag{9}$$

On the other hand, choose η_j, $x_{j-1} < \eta_j < x_j$, so that

$$f(\eta_j) - m_j < \frac{\varepsilon}{2(b - a)}$$

and we get

$$\sum_{j=1}^{n} f(\eta_j)(x_j - x_{j-1}) - \underline{S}(P; f) = \sum_{j=1}^{n} [f(\eta_j) - m_j](x_j - x_{j-1}) < \frac{\varepsilon}{2}. \tag{10}$$

From (8) and (9) and the definition of $\int_{a}^{b} f(x)\,dx$ we get

$$\int_{a}^{\bar{b}} f(x)\,dx - \int_{a}^{b} f(x)\,dx < \varepsilon \tag{11}$$

and from (8) and (10) and the definition of $\int_{a}^{b} f(x)\,dx$,

$$\int_{a}^{b} f(x)\,dx - \int_{\underline{a}}^{b} f(x)\,dx < \varepsilon \tag{12}$$

which implies that

$$\underline{\int_a^b} f(x)\,dx = \overline{\int_a^b} f(x)\,dx = \int_a^b f(x)\,dx.$$

Conversely, suppose that the lower and upper Darboux integrals exist and are equal to L. For every $\varepsilon > 0$ there correspond two partitions P_1 and P_2 such that

$$L - \underline{S}(P_1; f) < \varepsilon, \qquad \overline{S}(P_2; f) - L < \varepsilon,$$

by (5) and (6). Let $P_\varepsilon = P_1 \cup P_2$. Then relation (4) gives

$$L - \underline{S}(P_\varepsilon; f) < \varepsilon, \qquad \overline{S}(P_\varepsilon; f) - L < \varepsilon. \tag{13}$$

Let $P \supset P_\varepsilon$. Then by (4), (7), and (13),

$$-\varepsilon < \underline{S}(P_\varepsilon; f) - L \le \underline{S}(P; f) - L \le S(P; f) - L \le \overline{S}(P; f) - L$$

$$\le \overline{S}(P_\varepsilon; f) - L < \varepsilon.$$

This shows that

$$|S(P; f) - L| < \varepsilon$$

for any possible Cauchy sum $S(P; f)$ relative to P. Therefore f is Riemann integrable on $[a, b]$, and its integral is equal to L. $\qquad\square$

3.2. Proposition. f *is Riemann integrable on* $[a, b]$ *if and only if for every* $\varepsilon > 0$ *there is a partition* P *such that*

$$0 \le \overline{S}(P; f) - \underline{S}(P; f) < \varepsilon.$$

This proposition is immediate from Theorem 3.1. The proof is left for the reader (see Exercise 3G).

EXERCISES 3

A. Show that for a bounded function the upper and lower Darboux integrals always exist.

B. Show that if f is Riemann integrable on $[a, b]$, f is bounded.

C. Show that the Dirichlet function is not Riemann integrable.

D. Prove that every continuous function on $[a, b]$ is Riemann integrable on $[a, b]$.

E. Prove that a function which is Riemann integrable on a closed interval is integrable on any closed subinterval.

F. Prove that if f is Riemann integrable on $[a, b]$, then so is $|f|$. Give an example to show that the converse implication is false.

G. Prove Proposition 3.2.

H. Let $f: [0, 1] \to \mathbb{R}$ be defined by

$$f(x) = \begin{cases} 1/q & \text{if } x = p/q, \text{ where } p \text{ and } q \text{ are in } \mathbb{N} \ (q \neq 0) \text{ and have no common factors,} \\ 0 & \text{if } x \text{ is irrational, or } x = 0, 1. \end{cases}$$

(a) Prove that f is continuous at every irrational point of $[0, 1]$ and discontinuous where $f(x) \neq 0$.

(b) Prove that, in spite of having infinitely many discontinuities, f is Riemann integrable on $[0, 1]$, and $\int_0^1 f = 0$. (*Hint:* Sketch the graph. It looks like a Christmas tree.)

I. Let f be the Christmas tree function in Exercise H. Define $g: [0, 1] \to \mathbb{R}$ by $g(x) = 1$ if $0 < x \leq 1$ and $g(0) = 0$. Then both f and g are Riemann integrable. Show that the composite function $h = g \circ f$ is not Riemann integrable.

J. We define the *oscillation* ω_k of f over $[x_{k-1}, x_k]$ by

$$\omega_k = M_k - m_k,$$

where

$$M_k = \sup\{f(x): x_{k-1} \leq x \leq x_k\},$$
$$m_k = \inf\{f(x): x_{k-1} \leq x \leq x_k\}.$$

Theorem (Riemann, 1854). *A necessary and sufficient condition for the Riemann integrability of a bounded function f over $[a, b]$ is that if $\varepsilon > 0$ and $\delta > 0$, then there exists a partition*

$$P: a = x_0 < x_1 < \cdots < x_n = b,$$

such that the total length of the subintervals $[x_{k-1}, x_k]$ for which the oscillation ω_k is greater than ε is less than δ.

§4. Sets of Measure Zero

The purpose of this section is to set up the machinery which will have great intrinsic interest in every later part of the book.

It is easy to see that continuous and piecewise continuous functions are all Riemann integrable. For a piecewise continuous function the set of discontinuities is finite, i.e., the function is continuous everywhere except at finitely many points. What can we say about a function continuous everywhere except at infinitely many points? The Dirichlet functon, defined by $f(x) = 1$ for x rational and $f(x) = 0$ for x irrational, is not Riemann integrable on any interval $[a, b]$, since the lower Darboux integral is 0 and the upper Darboux integral is $b - a$ over the interval $[a, b]$, $a \neq b$. The set of discontinuities of this function is the interval $[a, b]$ for any a and b. What if the set of discontinuities is countably infinite? We shall see in §5 that a function having countably many discontinuities, or more generally, a set of discontinuities of measure zero, is Riemann integrable.

If I is a bounded interval with end points a and b, we define the *length* $|I|$ of I by $|I| = |b - a|$. Conventionally, we let $|\emptyset| = 0$.

4.1. Definition. A subset A of $\mathbb{R}$ is said to be a *set of measure zero* if for any $\varepsilon > 0$ there exists a sequence of bounded open intervals $I_1, I_2, \ldots$ such that:

(i) $A \subset \bigcup_{n=1}^{\infty} I_n$, and

(ii) $\sum_{n=1}^{\infty} |I_n| \leq \varepsilon$.

We often say that A *has measure zero* if A is a set of measure zero.

We observe that *in the preceding definition we can replace the open intervals by closed or half-open intervals*. In fact, if there exists a cover of A by a sequence of intervals (open or closed or half-open) of overall length $< \varepsilon$, then we replace the nth interval by another interval containing it, of length exceeding that of the nth interval by at most $\varepsilon/2^n$. Then A is covered by this new sequence of intervals, the overall length of which is less than 2ε. Since ε is arbitrary, so is 2ε.

No closed interval $[a, b]$, $a \neq b$, has measure zero. For if $[a, b]$ is covered by countably many open intervals, then it can be covered by a finite number of intervals extracted from the given covering because $[a, b]$ is compact. The sum of the lengths of just these open intervals clearly exceeds $b - a$, the length of the entire closed interval.

4.2. Proposition. *Any countable set (finite or infinite) has measure zero.*

Proof. Let $\{x_1, x_2, \ldots\}$ be a countable set and $\varepsilon > 0$. Each point x_n can be enclosed in an open interval of length $\varepsilon/2^n$, and $\sum_{n=1}^{\infty} \varepsilon/2^n = \varepsilon$. Therefore $\{x_1, x_2, \ldots\}$ has measure zero. $\qquad\square$

4.3. Proposition.

(a) *A subset of a set of measure zero has measure zero.*
(b) *If $A = \bigcup_{n=1}^{\infty} A_n$, where each A_n has measure zero, then A has measure zero.*

Proof. (a) is clear. We show (b). Let $\varepsilon > 0$. For each n, the set A_n can be covered by intervals I_{nk}, $k = 1, 2, \ldots$, where $\sum_{k=1}^{\infty} |I_{nk}| < \varepsilon/2^n$. Then the intervals I_{nk}, $n, k = 1, 2, \ldots$, satisfy

$$A = \bigcup_{n=1}^{\infty} A_n \subset \bigcup \{I_{nk} : n, k \in \mathbb{N}\},$$

$$\sum_{n=1}^{\infty} \sum_{k=1}^{\infty} |I_{nk}| < \sum_{n=1}^{\infty} \frac{\varepsilon}{2^n} = \varepsilon.$$

Consequently, A has measure zero. $\qquad\square$

The last proposition says that the union of a countable number of sets of measure zero is again a set of measure zero. The reader should not come to the conclusion that the sets of measure zero consist only of a countable number of points. Indeed, we have the following examples:

4.4. Example (The Cantor Ternary Set). One way of describing the Cantor ternary set F is as the set of real numbers in $[0, 1]$ which have a ternary expansion using only the digits 0 and 2. Equivalently, the Cantor set F is formed from the interval $[0, 1]$ by removing first the middle third, then the middle thirds of the remaining intervals, and so on indefinitely. It is understood that the intervals removed are open intervals. The set of points remaining after the infinite sequence of operations just described is the Cantor ternary set. To be more explicit, if we remove the open middle third of $[0, 1]$, we obtain the set $F_1 = [0, 1/3] \cup [2/3, 1]$. If we remove the open middle third of each of the two closed intervals in F_1, we obtain the set

$$F_2 = [0, 1/9] \cup [2/9, 1/3] \cup [2/3, 7/9] \cup [8/9, 1].$$

Hence F_2 is the union of 2^2 closed intervals of length $(1/3)^2$. Inductively, after F_n has been constructed and consists of the union of 2^n closed intervals of length $(1/3)^n$, we obtain F_{n+1} by removing the open middle third of each of these intervals. Then the Cantor ternary set is the intersection of the sets F_n (see Figure 1.2). Therefore the Cantor set F is closed. It is easy to show that F contains no interior points (Exercise 4C) and every point of the set is a cluster point of the set (Exercise 4A).

The Cantor ternary set has measure zero. The proof is rather easy. Let $\varepsilon > 0$. Then there exists n such that $(2/3)^n < \varepsilon$. Since $F \subset F_n$ and the overall sum of the lengths of intervals in $F_n = (2/3)^n$, F has measure zero.

To show that the Cantor ternary set is uncountable, we need the following characterization of the set:

4.5. Proposition. *Each point x of the Cantor ternary set F can be represented uniquely by a series of the form*

$$x = \sum_{n=1}^{\infty} \frac{a_n}{3^n}, \tag{1}$$

where each a_n is either 0 or 2, and every number thus represented is in F.

Figure 1.2

Proof. If we use the ternary system of place values, we can write the number x in (1) as

$$0.a_1 a_2 a_3 \ldots.$$

We now show that there cannot be more than one such representation. If

$$\sum_{n=1}^{\infty} \frac{a_n}{3^n} = \sum_{n=1}^{\infty} \frac{b_n}{3^n}, \tag{2}$$

where each b_n is also either 0 or 2, we must show that $a_n = b_n$ for every n. Suppose that $a_n \neq b_n$ for some n. Let p be the smallest natural number such that $a_p \neq b_p$. Then $|a_p - b_p| = 2$. Since $|a_n - b_n| \leq 2$ for each n, we have

$$0 = \left| \sum_{n=p}^{\infty} \frac{a_n - b_n}{3^n} \right| \geq \frac{1}{3^p} \left(|a_p - b_p| - \sum_{n=p+1}^{\infty} \frac{|a_n - b_n|}{3^n} \right),$$

$$\geq \frac{1}{3^p} \left(2 - \sum_{n=1}^{\infty} \frac{2}{3^n} \right) = \frac{1}{3^p}.$$

This is absurd. Therefore, $a_n = b_n$ for each n.

Let $G_{nk}, k = 1, 2, \ldots, 2^{n-1}$ be open intervals removed to obtain F_n (see Figure 1.2). Consider a number with a ternary representation $0.b_1 b_2 b_3 \ldots$, where each $b_i \in \{0, 1, 2\}$. It is easy to check that $0.b_1 b_2 b_3 \ldots \in G_{nk}$ for some k if and only if $b_m = 0$ or 2 for each $m < n$, and $b_n = 1$; for $m > n$, b_m has no restriction except that these b_m's are neither all 0's nor all 2's. (Examine the situation by denoting the endpoints of the open interval G_{nk} by their respective ternary representations.) This proves the proposition. $\qquad\square$

The Cantor ternary set is uncountable. This proof is also easy, being simply an application of Cantor's diagonal process. Suppose that $\{x_1, x_2, \ldots\}$ is a countable subset of F and let

$$x_1 = 0.a_{11} a_{12} a_{13} \ldots,$$

$$x_2 = 0.a_{21} a_{22} a_{23} \ldots,$$

$$x_3 = 0.a_{31} a_{32} a_{33} \ldots,$$

$$\ldots\ldots\ldots\ldots\ldots\ldots$$

be ternary represent tions with $a_{ij} = 0$ or 2. Now define

$$a_n = \begin{cases} 0 & \text{if } a_{nn} = 2, \\ 2 & \text{if } a_{nn} = 0. \end{cases}$$

Then the number $x = 0.a_1 a_2 a_3 \ldots$ is clearly in F, but it is not in the above list. Therefore any countable subset of F will omit at least one real number in F. This shows that F is uncountable. $\qquad\square$

Cantor introduced the ternary set in "De la puissance des ensembles parfaits de points" (1884).

4.6. Example (The Cantor n-Ary Set). The assertion that the Cantor ternary set has measure zero is equivalent to the fact that the set consists of points having a ternary expansion using only digits 0 and 2. In general, we have the *Cantor n-ary set*. Let n be a natural number, and let k be a natural number such that $0 < k < n - 1$. The Cantor n-ary set is formed by points in $[0, 1]$ whose n-ary expansion, written in the form

$$\frac{a_1}{n} + \frac{a_2}{n^2} + \frac{a_3}{n^3} + \cdots,$$

where $a_1, a_2, \ldots$ are natural numbers between 0 and $n - 1$, inclusive, and are different from k. Then it is easy to see that the set has measure zero. The proof is left to the reader as an exercise (Exercise 4D).

4.7. Remark. Notice that in the construction of the Cantor n-ary set we have actually used the following method. Let (I_n) be a sequence of disjoint open intervals in $[0, 1]$ of overall length 1, i.e., $\sum_{n=1}^{\infty} |I_n| = 1$, and let $K = [0, 1] \backslash \bigcup_{n=1}^{\infty} I_n$. Then K is of measure zero.

4.8. Example (A Generalized Cantor Set Which Is Not of Measure Zero). We modify Cantor's construction of the ternary set as follows: Let (a_n) be a sequence of positive real numbers such that $\sum_{n=1}^{\infty} a_n = \varepsilon < 1$. Now we imitate the construction of the Cantor ternary set F described in Example 4.4. Remove from the center of the unit interval $[0, 1]$ an open interval of length a_1 and obtain two closed intervals. From the center of each of these two intervals, remove an open interval of length $a_2/2$ to create 2^2 closed intervals. Now again from the center of each of these 2^2 intervals we remove an open interval of length $a_3/2^2$ to obtain 2^3 closed intervals. Repeating this process inductively, we start with 2^{n-1} closed intervals in the nth stage, and from the center of each of these intervals we remove an open interval of length $a_n/2^{n-1}$. Then the resulting set E is closed and does not contain any open interval. E cannot be of measure zero since if it were covered by a countable set of intervals of total length less than $1 - \varepsilon$, we should have the unit interval covered by a set of intervals of total length less than 1, which is absurd.

There is a common terminology involving sets of measure zero. A property which holds except on a set of measure zero is said to hold *almost everywhere*. For example, Dirichlet's function f is equal to zero almost everywhere. This is usually written $f(x) = 0$ a.e. The terminology "almost everywhere" may be modified in various ways; for instance, *almost all points* in a set have a certain property if all except those in a set of measure zero have it.

EXERCISES 4

A. Show that every point in the Cantor ternary set F is a cluster point of F.

B. Show that $\frac{1}{4}$ belongs to the Cantor ternary set F. (*Hint:* Ternary expansion.)

C. Show that the Cantor ternary set does not contain any open interval.

D. Show that the Cantor n-ary set has measure zero.

E. Show that if $A \subset \mathbb{R}$ does not have measure zero and $B \subset A$ has measure zero, then $A \backslash B$ does not have measure zero.

F. Show that the Dirichlet function defined on $[a, b]$ is equal to a Riemann integrable function on $[a, b]$ almost everywhere.

G. Let F be the Cantor ternary set. If $x \in F$ and

$$x = \sum_{n=1}^{\infty} \frac{a_n}{3^n},$$

then define $f: F \to \mathbb{R}$ by

$$f(x) = \sum_{n=1}^{\infty} \frac{a_n}{2^{n+1}}.$$

(a) Show that f is a function from F onto $[0, 1]$, where each point of $[0, 1]$ is in its binary representation.
(b) Show that f is continuous. Note that this provides another proof of the fact that F is uncountable.
(c) Show that f is monotone increasing; i.e., if $x_1 < x_2$, then $f(x_1) \le f(x_2)$.
(d) Show that f may be extended to a function $\tilde{f}$ which has domain $[0, 1]$, is monotone, nondecreasing, continuous, and is constant on each middle third interval removed in the way of the construction of F. This function is called *Lebesgue's singular function* (see Example 4.9, Chapter V). A way of describing $\tilde{f}$ is as follows: If

$$x = \sum_{n=1}^{\infty} \frac{a_n}{3^n}.$$

let us set

$$n = n(x) = \begin{cases} \min\{k: a_k = 1\} & \text{if } x \notin F, \\ \infty & \text{if } x \in F. \end{cases}$$

Then

$$\tilde{f}(x) = \sum_{k=1}^{n-1} \frac{a_k}{2^{k+1}} + \frac{1}{2^n}.$$

§5. Existence of the Riemann Integral

In §3 we gave a necessary and sufficient condition for a function defined on $[a, b]$ to be Riemann integrable (see Theorem 3.1 and Exercise 3C). In this section we shall put the condition into a more compact form in terms of the concept of measure zero. A close examination reveals that the condition given by Riemann (see Exercise 3C) states that the set of points of discontinuity of the function is to be of measure zero. Of course, Riemann himself did not regard his condition in terms of this concept, since it was about 50 years later that such a concept was explicitly introduced.

Inspired by Riemann's original proof of Riemann integrability, Lebesgue was able to show the following elegant and complete characterization of Riemann integrable functions.

5.1. Theorem (Lebesgue, 1902). *A bounded function f defined on $[a, b]$ is Riemann integrable if and only if it is continuous almost everywhere.*

Before we prove this theorem it will be convenient to introduce the following concepts:

5.2. Definition. Let $f: [a, b] \to \mathbb{R}$ and $c \in [a, b]$. The *limit superior, limit inferior,* and *oscillation* of f at c are, respectively, defined by

$$\overline{\lim_{x \to c}} f(x) = \lim_{\delta \to 0} \sup\{f(x): x \in [a, b] \cap [c - \delta, c + \delta]\},$$

$$\underline{\lim_{x \to c}} f(x) = \lim_{\delta \to 0} \inf\{f(x): x \in [a, b] \cap [c - \delta, c + \delta]\},$$

$$\omega(f; c) = \overline{\lim_{x \to c}} f(x) - \underline{\lim_{x \to c}} f(x).$$

It is clear that $\omega(f; c) \geq 0$.

5.3. Proposition. *The function f is continuous at x if and only if $\omega(f; x) = 0$.*

Proof. Suppose f is continuous at x. Let $\varepsilon > 0$. There is a $\delta > 0$ such that for all y in $[a, b]$ with $|x - y| < \delta$ we have

$$|f(x) - f(y)| < \frac{\varepsilon}{2}.$$

It follows that $\sup f(z) < \varepsilon/2 + f(x)$ and $\inf f(z) > - \varepsilon/2 + f(x)$, where sup and inf are taken over $[a, b] \cap [x - \delta, x + \delta]$. Therefore $0 \leq \omega(f; x) < \varepsilon$. Since ε is arbitrary, we conclude that $\omega(f; x) = 0$.

Conversely, assume that $\omega(f; x) = 0$. For $\varepsilon > 0$, there is a $\delta > 0$ such that $\sup f(z) - \inf f(z) < \varepsilon$, where sup and inf are taken on $[a, b] \cap [x - \delta, x + \delta]$. Then $y \in [x - \delta, x + \delta]$ implies $|f(x) - f(y)| < \varepsilon$, so that f is continuous at x. $\qquad\square$

The next theorem may be viewed as a generalization of the theorem that a continuous function on a compact set is uniformly continuous, and indeed the proof is essentially the same as the proof of the latter fact (see Theorem 6.6, Chapter Zero).

5.4. Proposition. *If $\omega(f; c) < \varepsilon$ for all c on the interval $[a, b]$, there exists $\delta > 0$ such that for all $x, y \in [a, b]$ with $|x - y| < \delta$ we have*

$$|f(x) - f(y)| < \varepsilon.$$

Proof. For every c in $[a, b]$ there exists $\delta_c > 0$ such that $\sup f(x) - \inf f(x) <$ ε, where sup and inf are taken over $[a, b] \cap [c - 2\delta_c, c + 2\delta_c]$. Thus, if $x, y \in$ $[a, b] \cap [c - 2\delta_c, c + 2\delta_c]$, then

$$|f(x) - f(y)| < \varepsilon. \tag{*}$$

Cover $[a, b]$ with a finite subset of the family of intervals $(c - \delta_c, c + \delta_c)$, say, $(c_k - \delta_k, c_k + \delta_k)$, where $k = 1, 2, \ldots, n$ and $\delta_k = \delta_{c_k}$. This can be done because $[a, b]$ is a compact set. Set $\delta = \min\{\delta_1, \ldots, \delta_n\}$. If $x, y \in [a, b]$ with $|x - y| <$ δ, and $x \in (c_k - \delta_k, c_k + \delta_k)$, then $y \in (c_k - 2\delta_k, c_k + 2\delta_k)$. Thus it follows from (*) that $|f(x) - f(y)| < \varepsilon$. $\qquad\square$

Notice that the same proof will work for a compact set instead of a closed interval (see Exercise 5B). We are now ready to prove the main theorem of this section.

Proof of Theorem 5.1. Let $D_n = \{x \in [a, b]: \omega(f; x) \geq 1/n\}$, and set $D = \bigcup_{n=1}^{\infty} D_n$. Then D is the set of all discontinuities of f in $[a, b]$.

We show that each D_m is of measure zero if f is integrable. Let m be a fixed number. If f is integrable, then for any $\varepsilon > 0$ there is a partition $P: a = x_0 <$ $x_1 < \cdots < x_n = b$ such that $\bar{S}(P; f) - \underline{S}(P; f) < \varepsilon/m$. Then the intervals (x_0, x_1), $\ldots, (x_{n-1}, x_n)$ can be split into two groups, where the intervals in the first group meet D_m, and those of the second group do not meet D_m. Then

$$\bar{S}(P; f) - \underline{S}(P; f) = \sum_1 (M_j - m_j)(x_j - x_{j-1})$$

$$+ \sum_2 (M_j - m_j)(x_j - x_{j-1}) < \frac{\varepsilon}{m},$$

where $\sum_1$ indicates summation over intervals of the first group and $\sum_2$ over the rest. On the intervals in the first group, $M_j - m_j \geq 1/m$, so that

$$\frac{1}{m} \sum_1 (x_j - x_{j-1}) \leq \sum_1 (M_j - m_j)(x_j - x_{j-1}) < \frac{\varepsilon}{m}.$$

Therefore, $\sum_1 (x_j - x_{j-1}) < \varepsilon$. This shows that the sum of the lengths of the intervals in the first group is less than ε. At the same time these intervals cover D_m. Therefore, D_m is of measure zero. It now follows from Proposition 4.3 that D has measure zero.

Conversely, assume that D has measure zero, and hence, each D_m has measure zero.

We claim that D_m is compact. Since D_m is a subset of the compact set $[a, b]$, it is enough to show that D_m is closed in $[a, b]$, or equivalently, that $[a, b] \backslash D_m$ is relatively open in $[a, b]$. Suppose $t \in [a, b] \backslash D_m$; then $\omega(f; t) < 1/m$. Now, from the definition of the limit superior and limit inferior, for $\varepsilon > 0$ there are $\delta_1 > 0, \delta_2 > 0$ such that, if $z \in [a, b] \cap (t - \delta_1, t + \delta_1)$, we have

$$f(z) < \overline{\lim_{x \to t}} f(x) + \frac{\varepsilon}{2},$$

and if $z \in [a, b] \cap (t - \delta_2, t + \delta_2)$, then

$$f(z) > \varliminf_{x \to t} f(x) - \frac{\varepsilon}{2}.$$

Let $\delta = \min\{\delta_1, \delta_2\}$. If $z \in [a, b] \cap (t - \delta, t + \delta)$, then we have

$$\varliminf_{x \to t} f(x) - \frac{\varepsilon}{2} < f(z) < \varlimsup_{x \to t} f(x) + \frac{\varepsilon}{2}.$$

This means, in particular, that for each $y \in [a, b] \cap (t - \delta, t + \delta)$

$$\varliminf_{x \to t} f(x) - \frac{\varepsilon}{2} \le \varlimsup_{z \to y} f(z) \le \varlimsup_{x \to t} f(x) + \frac{\varepsilon}{2},$$

$$\varliminf_{x \to t} f(x) - \frac{\varepsilon}{2} \le \varliminf_{z \to y} f(z) \le \varlimsup_{x \to t} f(x) + \frac{\varepsilon}{2}.$$

Therefore, if $y \in [a, b] \cap (t - \delta, t + \delta)$,

$$\omega(f; y) = \varlimsup_{z \to y} f(z) - \varliminf_{z \to y} f(z) \le \omega(f; t) + \varepsilon.$$

Take $\varepsilon < (1/m) - \omega(f; t)$; then $\omega(f; y) < 1/m$, and $y \in [a, b] \setminus D_m$. Consequently, $[a, b] \setminus D_m$ is open.

Since D_m is compact and of measure zero, there is a partition $P: a = x_0 < x_1 < \cdots < x_n = b$ such that $\sum_1 (x_j - x_{j-1}) < 1/m$, where $\sum_1$ indicates summation over intervals $[x_{j-1}, x_j]$ which meet D_m. Let K be the union of the intervals $[x_{j-1}, x_j]$ which do not meet D_m. Then if $x \in K$, $\omega(f; x) < 1/m$. Therefore by Proposition 5.4 there is $\delta > 0$ such that if $x, y \in K$ and $|x - y| < \delta$, $|f(x) - f(y)| < 1/m$. Let P' be a refinement of P whose norm $|P'| < \delta$; $P': a = y_0 < y_1 < \cdots < y_k = b$. Then

$$\bar{S}(P'; f) - \underline{S}(P'; f) = \sum_1 (M_j - m_j)(y_j - y_{j-1}) + \sum_2 (M_j - m_j)(y_j - y_{j-1})$$

$$< 2M \sum_1 (x_j - x_{j-1}) + \frac{b - a}{m}$$

$$< \frac{2M + b - a}{m},$$

where M is the maximum of $|M_j|$ and $|m_j|$ over $\sum_1$, and $\sum_1$ and $\sum_2$ are as before. This shows that f is integrable. $\square$

Theorem 5.1 has great intrinsic interest, since it clearly shows the scope of the Riemann integral. It can also be quite useful. For example, to show that if f and g are Riemann integrable then $f + g$ is also Riemann integrable, we simply observe that

$$D(f + g) \subset D(f) \cup D(g),$$

where $D(f)$ denotes the set of discontinuities of f, and we conclude $f + g$ is continuous almost everywhere.

Applying Theorem 5.1, we can easily construct a bounded function on $[a, b]$ which is not Riemann integrable.

5.5. Example. Let $A \subset [a, b]$ be such that $[a, b] \backslash A$ is dense in $[a, b]$ (i.e., every point of $[a, b]$ is a cluster point of $[a, b] \backslash A$) and has measure zero. If f and g are continuous on $[a, b]$ and $g \neq 0$, let the function h be defined by

$$h(x) = \begin{cases} f(x) & \text{for } x \in A, \\ f(x) + g(x) & \text{otherwise.} \end{cases}$$

Then h is everywhere discontinuous (see Exercise 5D). Therefore h is not Riemann integrable. The Dirichlet function is a special case of this kind. Later, we shall see that h is Lebesgue integrable.

EXERCISES 5

A. If $\lim_{x \to c} f(x)$ exists, show that

$$\lim_{x \to c} f(x) = \varliminf_{x \to c} f(x) = \varlimsup_{x \to c} f(x).$$

B. Prove Proposition 5.4 for a compact set.

C. If f and g are defined on $[a, b]$ and are Riemann integrable, show that the functions $\max\{f, g\}$, $\min\{f, g\}$, $|f|$, and f^2 are Riemann integrable.

D. Show that the function h defined in Example 5.5 is not Riemann integrable.

E. Let F be the Cantor ternary set. The function $\chi_F \colon [0, 1] \to \mathbb{R}$ which takes 1 on F and 0 otherwise is Riemann integrable. What is its integral over $[0, 1]$?

F. Let f be Riemann integrable on $[a, b]$ and $f(x) \geq 0$ for all x in $[a, b]$. Show that $\int_a^b f(x)\, dx = 0$ if and only if $f(x) = 0$ for almost all x.

§6. Deficiencies of the Riemann Integral

The definition of the integral by Cauchy and Riemann turns out to be inadequate from a more general point of view. The deficiencies of the Riemann integral can be roughly summed up in two brief statements. First, Riemann's definition has the drawback of applying only rarely; in other words, the class of all Riemann integrable functions is quite small as we have seen before in Example 5.5. Second and more seriously, limiting operations often lead to insurmountable difficulties. In fact, if $f_1, f_2, \ldots$ are each Riemann integrable on $[a, b]$ and $\lim f_n(x) = f(x)$ everywhere in $[a, b]$, then it is not in general true that

$$\lim \int_a^b f_n(x)\, dx = \int_a^b f(x)\, dx. \tag{$*$}$$

Three things may go wrong.

(1) The limit on the left side of (∗) may not exist.
(2) Even if this limit exists, the function f may not be Riemann integrable, and then the right side may be meaningless.
(3) Even if both sides exist, they may not be equal. Constructing an example corresponding to (1) is quite easy.

Now we give two examples illustrating the unpleasant possibilities (2) and (3).

6.1. Example. We enumerate the rationals in $[a, b]$ by $r_1, r_2, \ldots, r_n, \ldots$. Let

$$f_n(x) = \begin{cases} 1 & \text{if } x = r_k, k = 1, 2, \ldots, n, \\ 0 & \text{otherwise.} \end{cases}$$

Then f_n is Riemann integrable and $\int_a^b f_n(x)\, dx = 0$. On the other hand, $f(x) = \lim f_n(x)$ is the Dirichlet function which takes 1 on the rationals and 0 on the irrationals. Hence f is not Riemann integrable.

6.2. Example. Let

$$f_n(x) = -2n^2 x \exp(-n^2 x^2) + 2(n + 1)^2 x \exp[-(n + 1)^2 x^2].$$

Then the f_n's are telescoping and we have

$$\sum_{n=1}^{\infty} f_n(x) = -2x \exp(-x^2).$$

Now

$$\int_0^t [-2x \exp(-x^2)]\, dx = \exp(-t^2) - 1,$$

and

$$\sum_{n=1}^{\infty} \int_0^t f_n(x)\, dx = \sum_{n=1}^{\infty} [\exp(-n^2 t^2) - \exp(-(n + 1)^2 t^2)] = \exp(-t^2).$$

Therefore (∗) is not valid.

We remark further that the Riemann theory is not sufficient in describing a class of pairs of functions f and F satisfying the fundamental relation

$$F(x) - F(a) = \int_a^x f(x)\, dx \qquad \text{if and only if} \quad F'(x) = f(x).$$

We recall the following from calculus texts. Let there be defined on the closed interval $[a, b]$ a continuous function f whose derivative f' exists at every point of $[a, b]$ (at the endpoints a and b we consider one-sided derivatives). It is shown that if the derivative f' is Riemann integrable, then

$$f(x) - f(a) = \int_a^x f'(t)\, dt.$$

However, examples are known of derivatives which are bounded and not Riemann integrable; for intance, we have the following example:

6.3. Example. Let E be a generalized Cantor set in $[0, 1]$ which is not of measure zero (see Example 4.8). Suppose that the open interval (a, b) is removed from $[0, 1]$ to construct E. We define a function f_a by

$$f_a(x) = (x - a)^2 \sin \frac{1}{x - a}.$$

The derivative

$$f_a'(x) = 2(x - a) \sin \frac{1}{x - a} - \cos \frac{1}{x - a}$$

then vanishes at an infinite number of points in (a, b). Let c be such that

$$a + c = \sup\{x: a < x \le (a + b)/2, f_a'(x) = 0\}.$$

Define $F: [0, 1] \to \mathbb{R}$ such that $F(x) = 0$ at every point of E, and on each interval (a, b) removed from $[0, 1]$ to construct E we let

$$F(x) = \begin{cases} f_a(x) & \text{if } a < x \le a + c, \\ f_a(a + c) & \text{if } a + c \le x \le b - c, \\ -f_b(x) & \text{if } b - c \le x \le b. \end{cases}$$

Then F is continuous and differentiable everywhere on $[a, b]$ (see Figure 1.3). The derivative F' is bounded; in fact, $|F'(x)| \le 3$ since $|f_a'(x)| \le 3$. For $x \in E$, we obtain $F'(x) = 0$ from the definition of the derivative F' and the known limit $\lim_{y \to 0} y \sin(1/y) = 0$. If $x \in E$, then x is a cluster point of $[a, b]\backslash E$. Therefore, for every $x \in E$ and for any $\varepsilon > 0$ there is $y \notin E$ with $|x - y| < \varepsilon$ such that $|F'(y)| = 1$. (This claim can be justified by observing that if $\sin[1/(p - a)]$

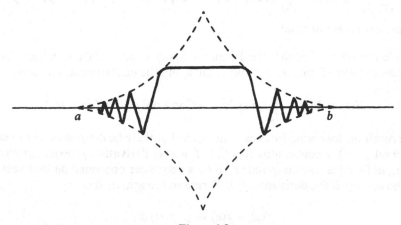

Figure 1.3

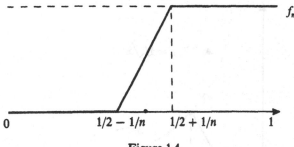

Figure 1.4

$= 0$, then $f_a'(p) = -\cos[1/(p - a)] = \pm 1$.) Hence F' cannot be continuous at any point of E. Also, the set of discontinuities of F' is E, which is not of measure zero. Therefore, by Theorem 5.1, F' is not Riemann integrable.

6.4. Example. We now consider the *space $C[a, b]$ of all real-valued continuous functions on $[a, b]$.* For f, g in $C[a, b]$, $|f - g| \in C[a, b]$, and we can define

$$d(f, g) = \int_a^b |f(x) - g(x)|\, dx.$$

Then d is a metric for $C[a, b]$. Unfortunately, *the space $C[a, b]$ with the metric d is not complete.* In fact, let

$$f_n(x) = \begin{cases} 0 & \text{for } 0 \leq x \leq \tfrac{1}{2} - 1/n, \\ (n/2)x + \tfrac{1}{2} - n/4 & \text{for } \tfrac{1}{2} - 1/n \leq x \leq \tfrac{1}{2} + 1/n, \\ 1 & \text{for } \tfrac{1}{2} + 1/n \leq x \leq 1. \end{cases}$$

(See Figure 1.4.) Then (f_n) is a Cauchy sequence in $C[0, 1]$ which converges to the function

$$f(x) = \begin{cases} 0 & \text{for } 0 \leq x \leq \tfrac{1}{2}, \\ 1 & \text{for } \tfrac{1}{2} < x \leq 1, \end{cases}$$

with respect to d; i.e., $\int_0^1 |f_n - f| \to 0$ as $n \to \infty$. But f is not in $C[0, 1]$. Therefore, $C[0, 1]$ is not complete with respect to the metric d.

A natural question arises: What space of functions can be obtained by adding all discontinuous functions which are limits of Cauchy sequences in $C[a, b]$ with respect to the metric d? This space will not be the space of all Riemann integrable functions. In fact, let

$$f_n(x) = \begin{cases} 0 & \text{for } 0 \leq x \leq 1/(n + 1), \\ n^{3/2}[(n + 1)x - 1] & \text{for } 1/(n + 1) \leq x \leq 1/n, \\ 1/\sqrt{x} & \text{for } 1/n \leq x \leq 1. \end{cases}$$

(See Figure 1.5.) Then (f_n) is a Cauchy sequence in $C[a, b]$ for the metric d which has no limit among Riemann integrable functions.

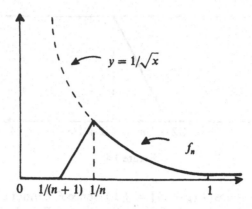

Figure 1.5

In other words, the space $R[a, b]$ *of all Riemann integrable functions on* $[a, b]$ is not complete, since the sequence in the last example is a Cauchy sequence in $R[a, b]$ which does not converge in $R[a, b]$. The fact that $R[a, b]$ is not complete is the most damaging piece of evidence in the case against the Riemann integral. The incompleteness of $R[a, b]$ has consequences which are almost as serious as those which would result from trying to develop real analysis without the completeness axiom of the real number system. In order to obtain a complete space generated by $C[a, b]$ with the limiting process as above, it is necessary to construct a new integral of greater scope than the Riemann integral.

We discussed here only a few of the deficiencies of the Riemann theory. It should be realized, however, that the deficiencies and inadequacies of the Riemann theory were not seriously regarded as problems at the time of the development of the theory. Inadequacies of the Riemann theory were mainly found by regarding the Riemann theory from the vantage point of Lebesgue's great discoveries.

EXERCISES 6

A. Find a sequence (f_n) of Riemann integrable functions on $[a, b]$ such that $\lim \int_a^b f_n(x)\, dx$ does not exist, but $\lim f_n = f$ a.e. for some bounded function f.

B. For $n \geq 2$, define $f_n: [0, 1] \to \mathbb{R}$ by

$$f_n(x) = \begin{cases} n^2 x & \text{for } 0 \leq x \leq 1/n, \\ -n^2(x - 2/n) & \text{for } 1/n \leq x \leq 2/n, \\ 0 & \text{for } 2/n \leq x \leq 1. \end{cases}$$

(See Figure 1.6.) Show that $f_n \to f$ pointwise on $[0, 1]$, where f is the zero function and

$$\lim \int_0^1 f_n(x)\, dx = 1 \neq 0 = \int_0^1 f(x)\, dx.$$

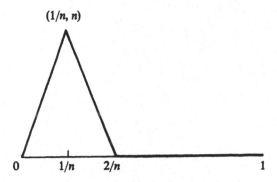

(1/n, n)

0   1/n   2/n   1

Figure 1.6

C. Let $f: [0, 1] \to \mathbf{R}$ be such that

$$f(x) = \begin{cases} x^2 \sin 1/x^2 & \text{for } x \neq 0, \\ 0 & \text{for } x = 0. \end{cases}$$

Then f is continuous and differentiable. Show that f' is not Riemann integrable.

CHAPTER II

The Lebesgue Integral: Riesz Method

Soon after Riemann's definition of the integral in 1854, its limitations became apparent. Numerous definitions of the integral for bounded as well as unbounded functions were successively proposed after 1854. At the beginning of this century, the French mathematician Henri Lebesgue (1875–1941) introduced in his doctoral dissertation at the Sorbonne, "Intégral, longueur, aire" (1902), a notion of the integral that was to become the keystone of modern analysis.

In Riemann integration, the domain over which the integral is taken is divided into a partition, and the integral is defined as the limit of the Cauchy sum for this partition as the norm of the partition diminishes. In Lebesgue integration, on the other hand, the domain over which the integral is taken is divided into a number of measurable sets. The integral is then defined as the limit of a certain sum taken for all these measurable sets as the number of measurable sets indefinitely increases. The distinction between the Lebesgue inegral and the Riemann integral rests essentially on the difference between the two methods of dividing the domain of integration into sets of points. We shall see this distinction in §10, Chapter III.

In this chapter, we present the scheme originally given by the Hungarian mathematician Frigyes Riesz (1880–1956) in "Sur l'intégrale de Lebesgue" (1920). Our most important source of inspiration has been the book of Frigyes Riesz and Bela Sz.-Nagy, *Functional Analysis* (1956). Our rationale is that Riesz's method is a better starting point for the account than that of Lebesgue, inasmuch as it is more economical, more direct, and leads more rapidly to the core of the subject. For example, we shall study the Lebesgue integral directly without appeal to measure theory and obtain at the outset in a very simple way the fundamental theorem of Lebesgue integration. This theorem states precise circumstances under which term-by-term integration is permissible.

This method requires only rudimentary theory of point sets. We shall see in Chapter III that the theory of measure follows as an application of the theory of integration.

We shall start with step functions and their integrals. The reason for this is to avoid assuming any knowledge of the theory of integration, including the Riemann theory. This chapter is therefore independent of the preceding chapter.

§1. Step Functions and Their Integrals

In this section we introduce the concept of a step function and define its integral in an obvious way. The reader may recall that the *characteristic function* χ_E of a set E is defined by

$$\chi_E(x) = \begin{cases} 1 & \text{for } x \in E, \\ 0 & \text{for } x \notin E. \end{cases}$$

1.1. Definition. A function $\varphi: [a, b] \to \mathbb{R}$ is called a *step function* if there exists a partition

$$a = x_0 < x_1 < \cdots < x_n = b$$

of the interval such that in every subinterval $I_k = (x_{k-1}, x_k)$ the function φ is constant; i.e., $\varphi(x) = a_k$ for $x \in I_k$ for $k = 1, 2, \ldots, n$ (see Figure 2.1).

Informally, a step function is one which assumes a finite number of values and assumes each of these values on an open interval. We can disregard the values of a step function at the partitioning points x_k or assign values to the function there arbitrarily, since they are finite in number (and therefore form a set of measure zero). It is permissible to omit finite sets and more generally, sets of measure zero in our considerations, since the values assumed by a function on a set of measure zero will not affect the value of the integral. For

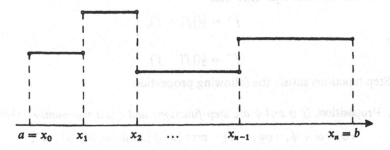

$a = x_0 \quad x_1 \quad x_2 \quad \cdots \quad x_{n-1} \quad x_n = b$

Figure 2.1

convenience we will represent the step function φ in the definition by

$$\varphi(x) = \sum_{k=1}^{n} a_k \chi_{I_k}(x).$$

We shall use the Greek letters $\varphi, \psi, \ldots$ for step functions.

Additional notations are useful at this point.

In the real number system every set of two elements a, b has both a unique least upper bound $\sup\{a, b\}$ and a unique greatest lower bound $\inf\{a, b\}$. These will be called the maximum and the minimum of $\{a, b\}$, respectively. Then

$$\sup\{a, b\} = \max\{a, b\} = \tfrac{1}{2}(a + b + |a - b|),$$

and

$$\inf\{a, b\} = \min\{a, b\} = \tfrac{1}{2}(a + b - |a - b|).$$

It follows immediately that if f and g are real-valued functions with common domain, then

$$\max\{f, g\} = \tfrac{1}{2}(f + g + |f - g|),$$

and

$$\min\{f, g\} = \tfrac{1}{2}(f + g - |f - g|).$$

are real-valued functions on the common domain whose values at any point x are $\max\{f(x), g(x)\}$ and $\min\{f(x), g(x)\}$, respectively (see Figure 2.2).

If f is a real-valued function, let f^+ and f^- be the nonnegative functions defined by

$$f^+ = \max\{f, 0\},$$

and

$$f^- = \max\{-f, 0\}.$$

The function f^+ is called the *positive part* of f, and f^- is called the *negative part* of f (see Figure 2.2). It is clear that

$$f = f^+ - f^-,$$

and

$$|f| = f^+ + f^-.$$

It follows from these equalities that

$$f^+ = \tfrac{1}{2}(|f| + f),$$

and

$$f^- = \tfrac{1}{2}(|f| - f).$$

Step functions satisfy the following properties:

1.2. Proposition. *If φ and ψ are step functions and c is a real number, then*

$$c\varphi, \quad \varphi + \psi, \quad \varphi\psi, \quad |\varphi|, \quad \max\{\varphi, \psi\}, \min\{\varphi, \psi\}, \varphi^+, \varphi^-,$$

are also step functions. Furthermore, if ψ is nonzero except at the partitioning points, then φ/ψ is also a step function.

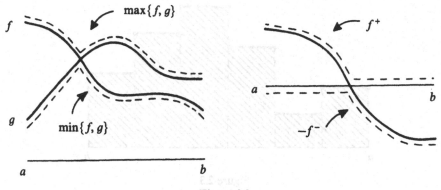

Figure 2.2

The proof is straightforward (see Exercise 1A).

We now define the integral for step functions in an obvious way and study its properties.

1.3. Definition. Let φ be a step function on $[a, b]$ represented by

$$\varphi(x) = \sum_{k=1}^{n} a_k \chi_{I_k}(x),$$

where $I_k = (x_{k-1}, x_k)$ for $1 \le k \le n$ and $a = x_0 < x_1 < \cdots < x_n = b$. The *integral* of this function is naturally defined by

$$\int_a^b \varphi(x)\, dx = \sum_{k=1}^{n} a_k (x_k - x_{k-1}).$$

The reader should recognize that our definition of the integral of a step function is actually the Riemann integral. For a nonnegative step function φ this is just the area under the graph of φ (Figure 2.3).

This section ends with the following elementary properties of the integral of step functions:

1.4. Proposition. *For step functions* φ, ψ, *and a real number c we have:*

(a) $\int_a^b [\varphi(x) + \psi(x)]\, dx = \int_a^b \varphi(x)\, dx + \int_a^b \psi(x)\, dx$ *(additive)*;

(b) $\int_a^b c\varphi(x)\, dx = c \int_a^b \varphi(x)\, dx$ *(homogeneous)*; *and*

(c) *if* $\varphi \ge 0$ *then* $\int_a^b \varphi(x)\, dx \ge 0$ *(positive)*.

These properties are well known. As a corollary of property (c), we have:

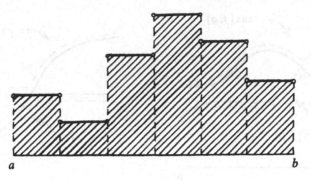

Figure 2.3

1.5. Corollary. *Let φ and ψ be step functions. Then:*

(a) $\int_a^b \varphi(x)\,dx \leq \int_a^b \psi(x)\,dx$ *if* $\varphi \leq \psi$;

(b) $\left| \int_a^b \varphi(x)\,dx \right| \leq \int_a^b |\varphi(x)|\,dx$;

(c) $\int_a^b \varphi^+(x)\,dx \leq \int_a^b |\varphi(x)|\,dx$; *and*

(d) $\int_a^b \varphi^-(x)\,dx \leq \int_a^b |\varphi(x)|\,dx$.

Exercises 1

A. Prove Proposition 1.2.

B. Let f be continuous on $[a, b]$. Show that the *greatest integer part* $[f]$ of the function f defined by

$$[f](x) = n \quad \text{if} \quad n \leq f(x) < n + 1$$

is a step function.

C. Show that:
   (a) $\min\{f, g\} = -(\max\{-f, -g\})$;
   (b) $\max\{f, g\} = -(\min\{-f, -g\})$;
   (c) $\max\{f, g\} + h = \max\{f + h, g + h\}$;
   (d) $\max\{f, g\} = f + g + \max\{-f, -g\}$;
   (e) $\min\{f, g\} + h = \min\{f + h, g + h\}$; and
   (f) $f + g = \max\{f, g\} + \min\{f, g\}$.

D. Prove Proposition 1.4.

E. Prove Corollary 1.5.

F. For any step function φ on $[a, b]$ and a real number h, we define φ_h by

$$\varphi_h(x) = \varphi(x + h).$$

Show that φ_h is a step function on $[a - h, b - h]$ and

$$\int_a^b \varphi(x)\,dx = \int_{a-h}^{b-h} \varphi_h(x)\,dx.$$

§2. Two Fundamental Lemmas

Our definitions of the Lebesgue inegral will be based on two lemmas in this section. Additional notions and notations are useful at this point.

2.1. Definition. The sequence (f_n) of functions defined on some common domain A of $\mathbb{R}$ is said to *converge almost everywhere* to the function f if

$$\lim f_n(x) = f(x)$$

for almost all $x \in A$; i.e., $\{x \in A: \lim f_n(x) \neq f(x)\}$ has measure zero. In this case we obtain write

$$\lim f_n = f \text{ a.e.} \qquad \text{or} \qquad f_n \to f \text{ a.e.}$$

It is obvious that ordinary convergence implies convergence almost everywhere.

The sequence of functions $f_n(x) = x^n$ defined on $[0, 1]$ converges to the function $f(x) = 0$ almost everywhere; more precisely, everywhere except at $x = 1$.

We also notice that convergence almost everywhere would not guarantee the uniqueness of the limit.

2.2. Proposition. *If* $\lim f_n = f$ *a.e., then* $\lim f_n = g$ *a.e. if and only if* $f = g$ *a.e.*

If (f_n) is a sequence of functions defined on some common domain, then $f_n \uparrow$ will indicate that, for each point x in that domain, the sequence $(f_n(x))$ of real numbers is monotone increasing. In this case, if we define $f(x) = \sup\{f_n(x): n \in \mathbb{N}\}$ for each point of the common domain, then f is a function defined on the same domain. We then write $f_n \uparrow f$. We notice that f is not, in general, real-valued, since the range of f may include the symbol ∞ if $(f_n(x))$ is not bounded above for some x. Similarly, we define $f_n \downarrow$ and $f_n \downarrow f$. These same notations will be used for a sequence (a_n) of real numbers. $a_n \uparrow$ will denote a monotone increasing sequence and $a_n \downarrow$ a monotone decreasing sequence.

In §1 we have studied some properties of the integral for step functions, in particular, additivity, homogeneity, and positivity of the inegral. Now we have the following continuity property of the integral which will play a fundamental role in our development of the Lebesgue integral.

2.3. First Fundamental Lemma. *Let* (φ_n) *be a monotone decreasing sequence of nonnegative step functions defined on* $[a, b]$. *Then* $\varphi_n \downarrow 0$ *almost everywhere on* $[a, b]$ *if and only if* $\lim \int_a^b \varphi_n(x) \, dx = 0$.

Proof. Suppose that $\varphi_n \downarrow 0$ almost everywhere on $[a, b]$. To show that $\int_a^b \varphi_n(x) \, dx \to 0$, let D_n be the set of discontinuities of φ_n for each n and let D_0 be the set outside of which (φ_n) converges to 0; i.e., $D_0 = \{x: \lim \varphi_n(x) \neq 0\}$. Then D_0 is of measure zero. Set $D = \bigcup_{n=0}^{\infty} D_n$. Then D has measure zero since it is a countable union of sets having measure zero. Therefore, for $\varepsilon > 0$, we can cover D with a sequence of open intervals I_n satisfying $\sum_{n=1}^{\infty} |I_n| < \varepsilon$. If $\xi \notin D$, then $\lim \varphi_n(\xi) = 0$, and hence there is a natural number $m = m(\xi)$ such that $\varphi_m(\xi) < \varepsilon$. Since φ_m is a step function, there is an open interval $I(\xi)$ containing ξ and on which φ_m is constant and equal to $\varphi_m(\xi)$. Notice that the closed interval $[a, b]$ is covered by (I_n) and $\{I(\xi): \xi \notin D\}$. Since $[a, b]$ is compact, we can extract finitely many open intervals $I_{n_1}, I_{n_2}, \ldots, I_{n_k}; I(\xi_1),$ $I(\xi_2), \ldots, I(\xi_q)$ from (I_n) and $\{I(\xi): \xi \notin D\}$, respectively, to cover $[a, b]$. Let $p = \max\{m(\xi_1), \ldots, m(\xi_q)\}$. If $r \geq p$ then $\varphi_r < \varepsilon$ on $A = I(\xi_1) \cup \cdots \cup I(\xi_q)$, since $\varphi_n \downarrow 0$ on the complement of D. Now, the set A can be expressed as the union of a finite number of disjoint intervals $I'_1, \ldots, I'_m$. Moreover, if $M = \sup\{\varphi_1(x): x \in [a, b]\}$, then $\varphi_r(x) \leq \varphi_1(x) \leq M$, and hence

$$\int_a^b \varphi_r(x)\,dx \leq M(|I_{n_1}| + \cdots + |I_{n_k}|) + \varepsilon(|I'_1| + \cdots + |I'_m|)$$

$$< M\varepsilon + \varepsilon(b - a) = \varepsilon(M + b - a).$$

Since ε is arbitrarily small, we then have $\int_a^b \varphi_r(x)\,dx \to 0$.

Conversely, let $\int_a^b \varphi_n(x)\,dx \to 0$. Since $\varphi_n \geq 0$ and $\varphi_n \downarrow$, then for each $x \in [a, b]$, $f(x) = \lim \varphi_n(x)$ exists. We now claim that f vanishes almost everywhere on $[a, b]$; i.e., the set $P = \{x \in [a, b]: f(x) > 0\}$ is of measure zero. Let

$$P_m = \{x \in [a, b]: f(x) \geq 1/m\}.$$

Then $P = \bigcup_{m=1}^{\infty} P_m$. To show that P is of measure zero, it is sufficient to prove that each P_m has measure zero.

Let n be fixed. Since $\varphi_n \geq f$, we have $\varphi_n(x) \geq 1/m$ for $x \in P_m$. We cover P_m with a finite number of intervals such that on each of these intervals φ_n is constant and $\geq 1/m$. The definition of a step function assures that this can be done. Let s_n be the overall length of these intervals. Then it is evident that

$$\int_a^b \varphi_n(x)\,dx \geq \frac{s_n}{m}.$$

Let $\varepsilon > 0$ be given. Since $\int_a^b \varphi_n(x)\,dx \to 0$ as $n \to \infty$, then for sufficiently large n, $\int_a^b \varphi_n(x)\,dx < \varepsilon/m$, and hence $s_n < \varepsilon$. This proves that P_m can be covered by a finite number of intervals of overall length less than ε. Therefore P_m is of measure zero.   □

2.4. Second Fundamental Lemma. *Let (φ_n) be a monotone increasing sequence of step functions defined on $[a, b]$. If there is a number A such that $\int_a^b \varphi_n(x)\,dx \leq A$ for all n, then the sequence (φ_n) converges almost everywhere on $[a, b]$.*

Proof. We may assume that all φ_n are nonnegative; otherwise we can simply consider the sequence $(\varphi_n - \varphi_1)$. Set $f(x) = \lim \varphi_n(x)$. To show that (φ_n) converges almost everywhere, it is sufficient to prove that the set $E = \{x \in [a, b]:$ $f(x) = \infty\}$ has measure zero. At each point x of E, we have $\lim \varphi_n(x) = \infty$. Since the set of discontinuities of all φ_n is of measure zero as it is a countable set, we can further assume that each φ_n is continuous at each point of E. Let $\varepsilon > 0$ be arbitrary, but otherwise fixed. Set $E_n = \{x \in [a, b]: \varphi_n(x) > A/\varepsilon\}$. Then $E \subset \bigcup_{n=1}^{\infty} E_n$. Since φ_n is a step function, E_n is the union of finitely many intervals. The total length s_n of the intervals in E_n is less than ε, since $As_n/\varepsilon \le \int_a^b \varphi_n(x)\, dx \le A$. Our aim is to show that $\bigcup_{n=1}^{\infty} E_n$ is covered by countably many intervals of overall length less than ε. Since $E_n \subset E_{n+1}$, the difference $E_{n+1} \backslash E_n$ may be expressed as the union of a finite number of disjoint intervals (why?). Since $\bigcup_{n=1}^{\infty} E_n = E_1 \cup (E_2 \backslash E_1) \cup \cdots$, it is now easy to see how to express $\bigcup_{n=1}^{\infty} E_n$ as a union of countably many disjoint intervals: first write down the intervals in E_1, then those of $E_2 \backslash E_1$, then those of $E_3 \backslash E_2$, and so on. If $E_{n+1} \backslash E_n = \varnothing$, we will consider the empty set as an interval for convenience. The first n intervals in this sequence will definitely be contained in E_n, and hence their overall length must be less than $s_n \le \varepsilon$. Therefore, the total length of the intervals in $\bigcup_{n=1}^{\infty} E_n$ will be less than ε. Since $\varepsilon > 0$ is arbitrary, this means that E is of measure zero. □

2.5. Corollary. *Let (φ_n) be a sequence of nonnegative step functions on $[a, b]$ such that $\sum_{n=1}^{\infty} \int_a^b \varphi_n(x)\, dx < \infty$. Then $\sum_{n=1}^{\infty} \varphi_n$ converges almost everywhere on $[a, b]$.*

The converse of the Second Fundamental Lemma, which states that a set of measure zero can be characterized by the property of the lemma, is also true.

2.6. Proposition. *A set $A \subset [a, b]$ is of measure zero if and only if there exists a monotone increasing sequence (φ_n) of step functions on $[a, b]$ such that $(\int_a^b \varphi_n(x)\, dx)$ converges and (φ_n) diverges on A.*

Proof. The "if" part follows from the Second Fundamental Lemma. We will prove the remaining part.

Suppose that A is a set of measure zero. Then for every natural number n there exsts a sequence (I_{nm}) of open intervals such that

$$A \subset \bigcup_{m=1}^{\infty} I_{nm} \quad \text{and} \quad \sum_{m=1}^{\infty} |I_{nm}| \le \frac{1}{2^n}.$$

We rearrange the double sequence (I_{nm}) into a single sequence (I_k) as we did in the proof of Proposition 3.1 of Chapter Zero. Then we have

$$\sum_{k=1}^{\infty} |I_k| = \sum_{n=1}^{\infty} \sum_{m=1}^{\infty} |I_{nm}| \le 1$$

and each $x \in A$ will be contained in I_k for infinitely many k's (why?). Let

$\varphi_n = \sum_{k=1}^n \chi_{I_k}$. Then it is clear that $\varphi_n \uparrow$ and (φ_n) diverges on A. However,

$$\int_a^b \varphi_n(x)\, dx \le \sum_{k=1}^n |I_k| \le 1$$

for all n. Thus (φ_n) is the required sequence.   □

EXERCISES 2

A. Prove Proposition 2.2.

B. Let (φ_n) be a sequence of step functions defined on $[a, b]$ and

$$M_n = \sup\{|\varphi_n(x)|: x \in [a, b]\}.$$

Show that if $M_n \to 0$, then $\int_a^b \varphi_n(x)\, dx \to 0$.

C. Prove Corollary 2.5.

D. Let f be continuous on $[a, b]$. Find a monotone increasing sequence (φ_n) of step functions on $[a, b]$ which converges to f almost everywhere on $[a, b]$ and such that $\lim \int_a^b \varphi_n < \infty$.

E. Construct a sequence (φ_n) of step functions on $[0, 1]$ such that $\varphi_n \uparrow$, and (φ_n) diverges on the Cantor ternary set and $(\int_0^1 \varphi_n(x)\, dx)$ converges.

§3. The Class L^+

In this and the next section we shall extend the integral to classes of functions larger than that of step functions by approximating more general functions by step functions. We shall first introduce a class of functions which are limits of step functions described in the Second Fundamental Lemma.

3.1. Definition. A function $f: [a, b] \to \mathbb{R}^*\ (= \mathbb{R} \cup \{\infty, -\infty\})$ belongs to the class L^+ if there is a monotone increasing sequence (φ_n) of step functions defined on $[a, b]$ such that:

(1) the sequence $(\int_a^b \varphi_n(x)\, dx)$ is bounded; and
(2) $f = \lim \varphi_n$ almost everywhere on $[a, b]$.

In view of the Second Fundamental Lemma, it is clear that if $f \in L^+$, then f is finite almost everywhere on $[a, b]$; i.e., $\{x \in [a, b]: f(x) = \pm\infty\}$ is of measure zero.

Let $f \in L^+$ and (φ_n) be a monotone increasing sequence of step functions which defines f in the sense of Definition 3.1. Then

$$\int_a^b \varphi_1(x)\, dx \le \int_a^b \varphi_2(x)\, dx \le \cdots \le \int_a^b \varphi_n(x)\, dx \le \cdots \le A$$

for some constant A. Therefore $\lim \int_a^b \varphi_n(x)\,dx$ exists as a finite number. This suggests that we define $\int_a^b f(x)\,dx$ as the limit of $\int_a^b \varphi_n(x)\,dx$.

3.2. Definition. If $f \in L^+$, we define the (*Lebesgue*) *integral* of f by the formula

$$\int_a^b f(x)\,dx = \lim \int_a^b \varphi_n(x)\,dx,$$

where (φ_n) has the same meaning as before.

But we still have to verify that the definition defines the integral uniquely; that is, that the integral is independent of the particular choice of the sequence (φ_n) which defines f. To show this, let us prove the following more general result.

3.3. Proposition. *Let f and g belong to the class L^+ and let (φ_n) and (ψ_n) be sequences of step functions which define f and g, respectively, in the sense of Definition 3.1. If $f \le g$ almost everywhere on $[a, b]$, then*

$$\lim \int_a^b \varphi_n(x)\,dx \le \lim \int_a^b \psi_n(x)\,dx.$$

Proof. Fix a number m and consider the monotone decreasing sequence $(\varphi_m - \psi_n)$. Then

$$\lim_{n\to\infty} (\varphi_m - \psi_n) = \varphi_m - \lim_{n\to\infty} \psi_n \le f - g \le 0$$

almost everywhere on $[a, b]$. Therefore the sequence of nonnegative functions $(\varphi_m - \psi_n)^+$ should be monotone decreasing to 0 almost everywhere on $[a, b]$. Applying the First Fundamental Lemma, we have

$$\lim \int_a^b (\varphi_m - \psi_n)^+(x)\,dx = 0.$$

Since $\varphi_m - \psi_n \le (\varphi_m - \psi_n)^+$ and

$$\int_a^b (\varphi_m - \psi_n)(x)\,dx \le \int_a^b (\varphi_m - \psi_n)^+(x)\,dx$$

we get

$$\int_a^b \varphi_m(x)\,dx \le \lim_{n\to\infty} \left[\int_a^b (\varphi_m - \psi_n)^+(x)\,dx + \int_a^b \psi_n(x)\,dx \right]$$

$$= \lim_{n\to\infty} \int_a^b \psi_n(x)\,dx.$$

Finally, letting $m \to \infty$, we obtain the desired inequality. $\square$

Proposition 3.3 actually shows that the integral of a function in the class L^+ is well defined. In fact, if $\varphi_n \uparrow f$ a.e. and $\psi_n \uparrow f$ a.e. in the sense of Definition 3.1, then we may take $g = f$ in Proposition 3.3 and deduce

$$\lim \int_a^b \varphi_n(x)\, dx = \lim \int_a^b \psi_n(x)\, dx.$$

3.4. Corollary. *If f and g are in the class L^+ and $f \leq g$ almost everywhere on $[a, b]$, then*

$$\int_a^b f(x)\, dx \leq \int_a^b g(x)\, dx.$$

3.5. Corollary. *If f and g are in the class L^+ and $f = g$ almost everywhere on $[a, b]$, then*

$$\int_a^b f(x)\, dx = \int_a^b g(x)\, dx.$$

We would like to see that our definition of the integral for the class L^+ has at least as broad a compass as that of Riemann. In fact, we have the following proposition:

3.6. Proposition. *Every Riemann integrable function on $[a, b]$ belongs to the class L^+ and its Riemann integral coincides with the integral for the class L^+.*

Proof. Let P_n be the partition of $[a, b]$:

$$P_n: \quad a = x_0 < x_1 < \cdots < x_{2^n} = b,$$

such that

$$x_k - x_{k-1} = \frac{b - a}{2^n}.$$

For this partition we associate a step function φ_n defined by

$$\varphi_n = \sum_{k=1}^{2^n} m_k \chi I_k,$$

where $m_k = \inf\{f(x): x \in I_k\}$ and $I_k = (x_{k-1}, x_k)$. Then $\varphi_n \uparrow f$ almost everywhere on $[a, b]$ (why?). Therefore $f \in L^+$ and

$$\lim \int_a^b \varphi_n(x)\, dx = \int_a^b f(x)\, dx$$

in the sense of the class L^+. On the other hand,

$$\lim \int_a^b \varphi_n(x)\, dx = \int_{\underline{a}}^b f(x)\, dx.$$

This proves that the Riemann integral is identical to the integral for the class L^+. $\square$

The following corollary is immediate:

3.7. Corollary. *If f is Riemann integrable on $[a, b]$, then both f and $-f$ belong to the class L^+.*

In view of Proposition 3.6, we can find the Lebesgue integral for a Riemann integrable function by applying the techniques we learned in calculus. It is now natural to ask whether the class L^+ is much larger than the class of all Riemann integrable functions. In fact, we have the following assertion:

3.8. Proposition. *The class L^+ is strictly bigger than that of all Riemann integrable functions:*

Proof. We know that the Dirichlet function $D(x)$, equal to 0 for x irrational and 1 for x rational, is not Riemann integrable. Therefore, it is sufficient to show that $D(x)$ is in the class L. But this is trivial since $D(x) = 0$ almost everywhere. □

The integral as defined for the class L^+ has the following basic properties:

3.9. Proposition. *For functions f, g in the class L^+ and a positive real number c, we have:*

(a) $\int_a^b [f(x) + g(x)] \, dx = \int_a^b f(x) \, dx + \int_a^b g(x) \, dx$;
(b) $\int_a^b cf(x) \, dx = c \int_a^b f(x) \, dx$; and
(c) *if $f \geq 0$ almost everywhere on $[a, b]$, then $\int_a^b f(x) \, dx \geq 0$.*

The proofs are easily carried over from Proposition 1.4 and Definition 3.2, and we take the liberty of leaving them to the reader.

One should notice that in the class L^+ it is not possible to subtract functions or to multiply them by negative numbers, since we are restricted to increasing sequences of step functions. In other words, the class L^+ is not a vector space. Therefore, it is natural to extend our class L^+ to a much wider class in which we an subtract functions and multiply them by any numbers. Such a class is introduced in the next section.

Before closing this section we shall construct functions f, g in the class L^+ for which $f - g$ does not belong to the class L^+.

3.10. Example. Let F be a generalized Cantor set (see Example 4.8, Chapter I) which is not of measure zero. Then the characteristic function χ_F of F is not a member of the class L^+ but $1 - \chi_F$ is a member of L^+.

Proof. Let (I_n) be a sequence of mutually disjoint open intervals removed from $[0, 1]$ to construct the generalized Cantor set F. For convenience we may assume that $\sum_{n=1}^{\infty} |I_n| = \frac{1}{2}$. We now show that $1 - \chi_F \in L^+$. In fact, let φ_n

be the step function defined by

$$\varphi_n(x) = \sum_{k=1}^{n} \chi_k(x),$$

where $\chi_k = \chi_{I_k}$. It is clear that $\varphi_n \uparrow 1 - \chi_F$ and $\int_0^1 \varphi_n(x)\, dx \le \frac{1}{2}$. Therefore, $1 - \chi_F$ belongs to the class L^+, and its integral is equal to $\frac{1}{2}$.

On the other hand, suppose $\chi_F \in L^+$. Then $\int_0^1 \chi_F(x)\, dx = \frac{1}{2}$ (why?). Let (φ_n) be a sequence of step functions such that $\varphi_n \ge 0$, $\varphi_n \uparrow \chi_F$, and

$$\lim \int_0^1 \varphi_n(x)\, dx = \int_0^1 \chi_F(x)\, dx.$$

Since the integral of χ_F is positive, there must exist a step function φ_n whose integral is positive. Let I be an open interval in some I_k on which φ_n is positive. Such an I can be easily found. (In fact, first find an open subinterval J on which φ_n is positive. If J is contained in some I_k, let $I = J$. Otherwise, we can find an I_k in J for large K since $|J| > 0$ (remember the construction of a generalized Cantor set). In this case, we let $I = I_k$.) Then $I \cap F = \varnothing$, and hence $\chi_F(x)\chi_I(x) = 0$ on $[0, 1]$. We now have

$$0 < \int_0^1 \varphi_n(x)\chi_I(x)\, dx \le \int_0^1 \chi_F(x)\chi_I(x)\, dx,$$

but the last integral is equal to 0. This shows that $0 < 0$, which is absurd. Therefore, we must conclude that $\chi_F \notin L^+$. $\square$

EXERCISES 3

A. If $f \in L^+$ and g is such that $g = f$ almost everywhere on $[a, b]$, show that $g \in L^+$.

B. Find a non-Riemann-integrable function which is equal to a Riemann integrable function almost everywhere.

C. Let f be a bounded function on $[a, b]$. For a partition P

$$a = x_0 < x_1 < \cdots < x_{2^n} = b,$$

where $x_k - x_{k-1} = (b - a)/2^n$, let

$$\varphi_n = \sum_{k=1}^{2^n} m_k \chi_{I_k} \quad \text{and} \quad \psi_n = \sum_{k=1}^{2^n} M_k \chi_{I_k},$$

where $m_k = \inf\{f(x): x \in I_k\}$, $M_k = \sup\{f(x): x \in I_k\}$, and $I_k = (x_{k-1}, x_k)$.
(1) If f is Riemann integrable on $[a, b]$, show that

$$\lim \varphi_n(x) = f(x) = \lim \psi_n(x)$$

almost everywhere.
(2) If $x \in (a, b)$ is not a partitioning point x_k for all partitions of the form above, then f is continuous at x if and only if $\lim \varphi_n(x)$ and $\lim \psi_n(x)$ exist and are equal.
(3) From (1) and (2) conclude that if f is Riemann integrable, then f is continuous almost everywhere.

D. For any $f \in L^+$ and a real number h, we define f_h by

$$f_h(x) = f(x + h).$$

Show that there is a monotone sequence (φ_n) of step functions on $[a - h, b - h]$ such that

$$\varphi_n \uparrow f_h \quad \text{and} \quad \int_a^b f(x)\,dx = \int_{a-h}^{b-h} f_h(x)\,dx.$$

Hint: Exercise 1F.

E. If $f \in L^+$ and $-f \in L^+$, show that there exists a Riemann integrable function g such that $f = g$ almost everywhere. (f need not be Riemann integrable.)

F. Prove Proposition 3.9.

G. If $f, g \in L^+$, show that $\max\{f, g\}, \min\{f, g\} \in L^+$.

H. Let F be a closed set in $[a, b]$. Show that the characteristic function of $[a, b]\backslash F$ belongs to the class L^+.

§4. The Lebesgue Integral

If $f, g \in L^+$, then $f - g$ does not necessarily belong to the class L^+. We have seen this in Example 3.10. In this section we shall complete the construction of the integral by extending it from the class L^+ to a bigger class in which it is possible to subtract functions or to multiply them by negative numbers.

4.1. Definition. By L we denote the class of functions which are differences of two L^+ functions. Thus if f and g are elements of L^+, then $f - g$ is an element of the class L.

The following proposition enables us to carry out all the natural functional operations in the class L:

4.2. Proposition. Let f and g be in the class L and let c be a real number. Then the functions

$$f + g, \quad cf, \quad |f|, \quad \max\{f, g\}, \quad \min\{f, g\}, \quad f^+, \quad f^-,$$

are all in the class L.

Proof. (a) Write $f = f_1 - f_2$ and $g = g_1 - g_2$, where $f_i \in L^+$ and $g_i \in L^+$, $i = 1, 2$. Then

$$f + g = (f_1 + g_1) - (f_2 + g_2)$$

and since $f_1 + g_1 \in L^+$ and $f_2 + g_2 \in L^+$, it follows that $f + g \in L^+$.

(b) If $c \geq 0$, then $f = f_1 - f_2, f_i \in L^+$, implies that $cf = cf_1 - cf_2, cf_i \in L^+$, and hence $cf \in L$. If $c < 0$, then $-c > 0$ and $cf = (-c)f_2 - (-c)f_1$ shows that $cf \in L$.

(c) Let $f = f_1 - f_2, f_i \in L^+$. Then $\max\{f_1, f_2\} \in L^+$ and $\min\{f_1, f_2\} \in L^+$ (see Exercise 3G). Therefore,

$$|f| = \max\{f_1, f_2\} - \min\{f_1, f_2\}$$

belongs to the class L.

(d) Since

$$\max\{f, g\} = \tfrac{1}{2}(f + g + |f - g|)$$

and

$$\min\{f, g\} = \tfrac{1}{2}(f + g - |f - g|)$$

both $\max\{f, g\}$ and $\min\{f, g\}$ belongs to the class L by (a), (b), and (c) above. In particular, we have f^+ and f^- in L.   □

We now give the definition of the integral for functions in the class L.

4.3. Definition. Let $f \in L$ be such that

$$f = f_1 - f_2 \qquad \text{where} \qquad f_1, f_2 \in L^+.$$

The *integral* of f is defined as follows:

$$\int_a^b f(x)\, dx = \int_a^b f_1(x)\, dx - \int_a^b f_2(x)\, dx.$$

We must show that this definition is also unique. This can be shown very simply. In fact, if

$$f = f_1 - f_2 = g_1 - g_2$$

almost everywhere on $[a, b]$, where $f_i, g_i \in L^+$, $i = 1, 2$, then it follows that

$$f_1 + g_2 = f_2 + g_1.$$

Since the additivity for integrals in the class L^+ has already been established (see Proposition 3.9),

$$\int_a^b f_1(x)\, dx + \int_a^b g_2(x)\, dx = \int_a^b g_1(x)\, dx + \int_a^b f_2(x)\, dx$$

holds; that is,

$$\int_a^b f_1(x)\, dx - \int_a^b f_2(x)\, dx = \int_a^b g_1(x)\, dx - \int_a^b g_2(x)\, dx$$

which was to be proved.

The class of functions L is called *the class of Lebesgue integrable functions*, and the integral for the class L is called *the Lebesgue integral*. It is clear that $L^+ \subset L$ and the inclusion is strict, as shown by Example 3.10.

In the future we shall often say *integrable* instead of *Lebesgue integrable* for the sake of brevity. If the Riemann integral is in question, this will be explicitly stated.

4.4. Proposition. *For functions f, g in the class L and a real number c, we have:*

(a) $\int_a^b [f(x) + g(x)] \, dx = \int_a^b f(x) \, dx + \int_a^b g(x) \, dx$ *(additive);*

(b) $\int_a^b cf(x) \, dx = c \int_a^b f(x) \, dx$ *(homogeneous); and*

(c) *if* $f \geq 0$, *then* $\int_a^b f(x) \, dx \geq 0$ *(positive).*

Proof. (a) Let $f = f_1 - f_2$, $g = g_1 - g_2$, where f_i and g_i are in the class L^+, $i = 1, 2$. Then $f + g = (f_1 + g_1) - (f_2 + g_2)$, and by definition

$$\int_a^b [f(x) + g(x)] \, dx$$

$$= \int_a^b [f_1(x) + g_1(x)] \, dx - \int_a^b [f_2(x) + g_2(x)] \, dx,$$

$$= \int_a^b f_1(x) \, dx + \int_a^b g_1(x) \, dx - \int_a^b f_2(x) \, dx - \int_a^b g_2(x) \, dx,$$

$$= \left[\int_a^b f_1(x) \, dx - \int_a^b f_2(x) \right] dx + \left[\int_a^b g_1(x) \, dx - \int_a^b g_2(x) \right] dx,$$

$$= \int_a^b f(x) \, dx + \int_a^b g(x) \, dx.$$

(b) If $c \geq 0$, then $cf = cf_1 - cf_2$, $cf_i \in L^+$, and

$$\int_a^b cf(x) \, dx = \int_a^b cf_1(x) \, dx - \int_a^b cf_2(x) \, dx$$

$$= c \int_a^b f_1(x) \, dx - c \int_a^b f_2(x) \, dx$$

$$= c \left[\int_a^b f_1(x) \, dx - \int_a^b f_2(x) \, dx \right]$$

$$= c \int_a^b f(x) \, dx.$$

If $c < 0$, then $cf = (-c)f_2 - (-c)f_1$ with $(-c)f_i \in L^+$. Therefore,

$$\int_a^b cf(x) \, dx = \int_a^b (-c)f_2(x) \, dx - \int_a^b (-c)f_1(x) \, dx$$

$$= (-c) \int_a^b f_2(x) \, dx - (-c) \int_a^b f_1(x) \, dx$$

$$= c \left[\int_a^b f_1(x) \, dx - \int_a^b f_2(x) \, dx \right]$$

$$= c \int_a^b f(x) \, dx.$$

(c) Let $f = f_1 - f_2 \geq 0$, where $f_i \in L^+$. Then $f_1 \geq f_2$, and hence,

$$\int_a^b f_1(x)\, dx \geq \int_a^b f_2(x)\, dx$$

by Corollary 1.5. Therefore $\int_a^b f(x)\, dx \geq 0$.　　　　　　　　　□

The following proposition is fundamental:

4.5. Proposition. *If $f \in L$, then $|f| \in L$ and*

$$\left| \int_a^b f(x)\, dx \right| \leq \int_a^b |f(x)|\, dx.$$

Proof. By Proposition 4.2, we have $|f| \in L$. The integral inequality is a consequence of the positive property of the integral (see Proposition 4.4).　　□

We conclude this section with the following proposition, which is also of some interest:

4.6. Proposition. *If $f \in L$, then there exists a sequence (φ_n) of step functions on $[a, b]$ such that $\varphi_n \to f$ almost everywhere on $[a, b]$ and*

$$\lim \int_a^b \varphi_n(x)\, dx = \int_a^b f(x)\, dx.$$

Proof. Write $f = f_1 - f_2$, where $f_i \in L^+$, $i = 1, 2$. By Definition 3.1 of the class L^+, there exist two monotonic increasing sequences (ψ_n) and (ψ_n') of step functions on $[a, b]$ for which $\psi_n \to f_1$ and $\psi_n' \to f_2$ almost everywhere on $[a, b]$. Let

$$\varphi_n = \psi_n - \psi_n'.$$

Then (φ_n) is also a sequence of step functions which converges to f almost everywhere on $[a, b]$. Moreover,

$$0 \leq \left| \int_a^b f(x)\, dx - \int_a^b \varphi_n(x)\, dx \right| = \left| \int_a^b [f(x) - \varphi_n(x)]\, dx \right|$$

$$= \left| \int_a^b [f_1(x) - f_2(x) - \psi_n(x) + \psi_n'(x)]\, dx \right|$$

$$\leq \left| \int_a^b f_1(x)\, dx - \int_a^b \psi_n(x)\, dx \right| + \left| \int_a^b f_2(x)\, dx - \int_a^b \psi_n'(x)\, dx \right|.$$

The last two terms converge to zero as $n \to \infty$ by the definition of the integral for the class L^+. Therefore,

$$\lim \int_a^b \varphi_n(x)\, dx = \int_a^b f(x)\, dx.$$　　　　　□

EXERCISES 4

A. Show that, in the class L, if $f \leq g$, then $\int_a^b f(x)\, dx \leq \int_a^b g(x)\, dx$.

B. Show that for a Riemann integrable function f the converse of Proposition 4.5 is not valid. (Find a function f such that $|f|$ is Riemann integrable but f is not Riemann integrable.)

C. Let $f \in L$ and $\varepsilon > 0$. Show that there exists a step function φ on $[a, b]$ such that

$$\int_a^b |f(x) - \varphi(x)|\, dx < \varepsilon.$$

§5. The Beppo Levi Theorem—Monotone Convergence Theorem

As is well known, the limit of a sequence of functions that are integrable in the sense of Riemann is not necessarily Riemann integrable, even if the sequence is bounded and all functions in the sequence have bounded integrals (see Example 6.1, Chapter I). This is a major drawback of the Riemann theory of integration, apart from the fact that even relatively simple functions are not integrable in the Riemann sense. Some of the difficulties occurring in the integration of sequences can be overcome by introducing the Lebesgue integral. In this and the next section we shall study the behavior of the Lebesgue integral in limiting processes.

We shall begin by considering the class L^+. Recall that a function in the class L^+ is an almost everywhere limit of a monotone increasing sequence of step functions on $[a, b]$ whose integrals have a common bound (see Definition 3.1). We have seen that the class L^+ is much larger than that of step functions. A natural question to ask is: Can we get a larger class of functions than the class L^+ by repeating the same limit passage through monotone increasing sequences of L^+ functions whose integrals have a common bound? The same question applies to the class L in place of L^+.

5.1. Proposition. *Let (f_n) be a monotone increasing sequence of functions in the class L^+ such that*

$$\int_a^b f_n(x)\, dx \leq A$$

for all n. Then (f_n) converges almost everywhere on $[a, b]$, and if $f = \lim f_n$ a.e., then $f \in L^+$ and

$$\int_a^b f(x)\, dx = \lim \int_a^b f_n(x)\, dx.$$

This means that the limit and the integral sign can be interchanged.

Proof. For each k let (φ_{kn}) be a monotone increasing sequence of step functions which converges to f_k almost everywhere on $[a, b]$ as shown in the following table.

$$\varphi_{11} \leq \varphi_{12} \leq \cdots \leq \varphi_{1n} \leq \cdots \quad \to f_1,$$

$$\varphi_{21} \leq \varphi_{22} \leq \cdots \leq \varphi_{2n} \leq \cdots \quad \to f_2,$$

$$\cdots\cdots\cdots\cdots\cdots\cdots\cdots\cdots\cdots$$

$$\varphi_{j1} \leq \varphi_{j2} \leq \cdots \leq \varphi_{jn} \leq \cdots \quad \to f_j,$$

$$\cdots\cdots\cdots\cdots\cdots\cdots\cdots\cdots\cdots \quad \vdots$$

Put $\varphi_n = \max\{\varphi_{jn}: 1 \leq j \leq n\}$. It is obvious that the step functions φ_n form a monotone increasing sequence. Since $\varphi_{jn} \leq f_j \leq f_n$ for $j \leq n$, we also have $\varphi_n \leq f_n$, and hence,

$$\int_a^b \varphi_n(x)\, dx \leq \int_a^b f_n(x)\, dx \leq A$$

for all n. By the Second Fundamental Lemma 2.4, (φ_n) converges to a function, say g, almost everywhere on $[a, b]$. From the definition of the class L^+, we have $g \in L^+$ and

$$\int_a^b g(x)\, dx = \lim \int_a^b \varphi_n(x)\, dx. \qquad (*)$$

On the other hand, $\varphi_{jn} \leq \varphi_n \leq f_n$ for any $j \leq n$, and if we fix j and take the limit as $n \to \infty$, we get $f_j \leq g \leq f$. Now letting $j \to \infty$, we have $f \leq g \leq f$, and hence $f = g$ almost everywhere. Therefore $f \in L^+$ and $\int_a^b f(x)\, dx = \int_a^b g(x)\, dx$. Since

$$\int_a^b \varphi_n(x)\, dx \leq \int_a^b f_n(x)\, dx \leq \int_a^b f(x)\, dx.$$

we conclude from $(*)$ that

$$\lim \int_a^b f_n(x)\, dx = \int_a^b f(x)\, dx. \qquad \square$$

5.2. Corollary. *For every series $\sum_{n=1}^\infty f_n$ in the class L^+ for which $f_n \geq 0$ and the integrals of partial sums of the series have a common bound, i.e.,*

$$\int_a^b \left[\sum_{n=1}^k f_n(x) \right] dx \leq A \qquad \text{for all } k,$$

the series $\sum_{n=1}^\infty f_n$ converges almost everywhere on $[a, b]$, and if $f = \sum_{n=1}^\infty f_n$ a.e., then $f \in L^+$ and

$$\int_a^b f(x)\, dx = \sum_{n=1}^\infty \int_a^b f_n(x)\, dx.$$

Proof. Put $g_k = \sum_{n=1}^k f_n$, and apply the preceding theorem to g_k. $\qquad \square$

The following two equivalent theorems are due to the Italian mathematician Beppo Levi, "Sopra l'integrazione delle series" (1906a). The second, Theorem 5.5, is often called the *monotone convergence theorem*, which is more suggestive than the Beppo Levi theorem; hence we will always refer to this theorem as the monotone convergence theorem.

5.3. The Beppo Levi Theorem. *For every series $\sum_{n=1}^{\infty} f_n$ of functions in the class L for which $f_n \geq 0$ and the integrals of partial sums have a common bound, i.e.,*

$$\int_a^b \left[\sum_{n=1}^{k} f_n(x) \right] dx \leq A \qquad \text{for all } k,$$

$\sum_{n=1}^{\infty} f_n$ converges almost everywhere on $[a, b]$, and if f is a function that is equal to $\sum_{n=1}^{\infty} f_n$ almost everywhere, then $f \in L$ and

$$\int_a^b f(x) \, dx = \sum_{n=1}^{\infty} \int_a^b f_n(x) \, dx.$$

We first prove the following lemma:

5.4. Lemma. *If $f \in L$ and $\varepsilon > 0$, then there exist $f_1, f_2 \in L^+$ such that $f = f_1 - f_2, f_2 \geq 0$, and $\int_a^b f_2(x) \, dx < \varepsilon$.*

Proof of Lemma. Let $f = g - h$, where $g, h \in L^+$. Choose a monotone increasing sequence (φ_n) of step functions for which $h = \lim \varphi_n$ a.e. and

$$\int_a^b h(x) \, dx = \lim \int_a^b \varphi_n(x) \, dx.$$

We write

$$f = g - h = (g - \varphi_n) - (h - \varphi_n).$$

Notice that both $g - \varphi_n$ and $h - \varphi_n$ are members of L^+, and that $h - \varphi_n \geq 0$. For $\varepsilon > 0$ there is a natural number n_0 such that

$$\int_a^b [h(x) - \varphi_{n_0}(x)] \, dx < \varepsilon$$

by the definition of the integral of h. Let $f_1 = g - \varphi_{n_0}$ and $f_2 = h - \varphi_{n_0}$; then f_1, f_2 satisfy the lemma. $\qquad \square$

Proof of Theorem 5.3. By the above lemma, for each n, find f_{n1} and f_{n2} in L^+ such that

$$f_n = f_{n1} - f_{n2}, \qquad f_{n2} \geq 0, \qquad \int_a^b f_{n2}(x) \, dx < \frac{1}{2^n}.$$

Then the series $\sum_{n=1}^{\infty} f_{n2}$ satisfies the condition of Corollary 5.2. Hence the series converges almost everywhere on $[a, b]$, and if we write

$$f_2 = \sum_{n=1}^{\infty} f_{n2},$$

then $f_2 \in L^+$ and

$$\int_a^b f_2(x) \, dx = \sum_{n=1}^{\infty} \int_a^b f_{n2}(x) \, dx.$$

We now show that $\sum_{n=1}^{\infty} f_{n1}$ satisfies the condition of Corollary 5.2. Since $f_n \geq 0$,

$$f_{n1} = f_n + f_{n2} \geq 0.$$

Furthermore, the integrals of partial sums of $\sum_{n=1}^{\infty} f_{n1}$ have a common bound. In fact,

$$\int_a^b \left[\sum_{n=1}^{k} f_{n1}(x) \right] dx = \int_a^b \left[\sum_{n=1}^{k} f_n(x) \right] dx + \int_a^b \left[\sum_{n=1}^{k} f_{n2}(x) \right] dx \leq A + 1.$$

Therefore, $\sum_{n=1}^{\infty} f_{n1}$ converges almost everywhere on $[a, b]$. Let $f_1 = \sum_{n=1}^{\infty} f_{n1}$. Then $f_1 \in L^+$ and

$$\int_a^b f_1(x) \, dx = \sum_{n=1}^{\infty} \int_a^b f_{n1}(x) \, dx.$$

It follows that

$$\sum_{n=1}^{\infty} f_n = \sum_{n=1}^{\infty} f_{n1} - \sum_{n=1}^{\infty} f_{n2} = f_1 - f_2$$

almost everywhere. Let $f = \sum_{n=1}^{\infty} f_n$. Then $f \in L$ and

$$\int_a^b f(x) \, dx = \int_a^b f_1(x) \, dx - \int_a^b f_2(x) \, dx$$

$$= \sum_{n=1}^{\infty} \int_a^b f_{n1}(x) \, dx - \sum_{n=1}^{\infty} \int_a^b f_{n2}(x) \, dx$$

$$= \sum_{n=1}^{\infty} \left[\int_a^b f_{n1}(x) \, dx - \int_a^b f_{n2}(x) \, dx \right]$$

$$= \sum_{n=1}^{\infty} \int_a^b f_n(x) \, dx. \qquad \qquad \Box$$

5.5. Monotone Convergence Theorem. *Let* (f_n) *be a monotone increasing sequence of functions in the class* L *whose integrals have a common bound, i.e.,*

$$\int_a^b f_n(x) \, dx \leq A \qquad for \ all \ n.$$

Then (f_n) *converges almost everywhere on* $[a, b]$, *and if* $f = \lim f_n$, *then* $f \in L$ *and*

$$\int_a^b f(x) \, dx = \lim \int_a^b f_n(x) \, dx.$$

Proof. This follows immediately from the Beppo Levi theorem. In fact,

$$f = \lim f_n = f_1 + \lim \sum_{k=1}^{n} (f_{k+1} - f_k)$$

$$= f_1 + \sum_{k=1}^{\infty} (f_{k+1} - f_k),$$

and hence

$$\int_a^b f(x)\, dx = \lim \int_a^b f_n(x)\, dx. \qquad \square$$

5.6. Corollary. *Let* (f_n) *be a monotone decreasing sequence of functions in the class L such that*

$$\int_a^b f_n(x)\, dx \geq A \qquad \text{for all } n.$$

Then (f_n) *converges almost everywhere on* $[a, b]$, *and if* $f = \lim f_n$, *then* $f \in L$ *and*

$$\int_a^b f(x)\, dx = \lim \int_a^b f_n(x)\, dx.$$

So far we have discussed the integration of series with nonnegative terms. But an immediate generalization of Beppo Levi's theorem for arbitrary functions in L can be given in the following form, which will also be called the Beppo Levi theorem.

5.7. Theorem. *Let* $\sum_{n=1}^{\infty} f_n$ *be a series of functions in the class L. Suppose that*

$$\sum_{n=1}^{\infty} \int_a^b |f_n(x)|\, dx$$

converges. Then the series $\sum_{n=1}^{\infty} f_n$ *converges almost everywhere on* $[a, b]$, *and if* $f = \sum_{n=1}^{\infty} f_n$, *then* $f \in L$ *and*

$$\int_a^b f(x)\, dx = \sum_{n=1}^{\infty} \int_a^b f_n(x)\, dx.$$

Proof. Since $|f_n| = f_n^+ + f_n^-$, we have

$$\sum_{n=1}^{\infty} \int_a^b f_n^+(x)\, dx < \infty \qquad \text{and} \qquad \sum_{n=1}^{\infty} \int_a^b f_n^-(x)\, dx < \infty.$$

Therefore, by the Beppo Levi theorem,

$$\sum_{n=1}^{\infty} f_n^+ \in L \qquad \text{and} \qquad \sum_{n=1}^{\infty} f_n^- \in L,$$

and

$$\int_a^b \left[\sum_{n=1}^\infty f_n^+(x) \right] dx = \sum_{n=1}^\infty \int_a^b f_n^+(x) \, dx,$$

$$\int_a^b \left[\sum_{n=1}^\infty f_n^-(x) \right] dx = \sum_{n=1}^\infty \int_a^b f_n^-(x) \, dx.$$

It follows that

$$\sum_{n=1}^\infty f_n = \sum_{n=1}^\infty (f_n^+ - f_n^-) = \sum_{n=1}^\infty f_n^+ - \sum_{n=1}^\infty f_n^- \in L,$$

and if $f = \sum_{n=1}^\infty f_n$, then

$$\int_a^b f(x) \, dx = \sum_{n=1}^\infty \int_a^b f_n^+(x) \, dx - \sum_{n=1}^\infty \int_a^b f_n^-(x) \, dx$$

$$= \sum_{n=1}^\infty \int_a^b (f_n^+ - f_n^-)(x) \, dx = \sum_{n=1}^\infty \int_a^b f_n(x) \, dx. \qquad \square$$

Many results follow from the Beppo Levi theorem (see Exercises 5A and 5B), but the following proposition is especially important:

5.8. Proposition. *Let $f \in L$. Then $\int_a^b |f(x)| \, dx = 0$ if and only if $f = 0$ almost everywhere on $[a, b]$.*

Proof. One direction needs no proof. Suppose that $\int_a^b |f(x)| \, dx = 0$. Let $f_n = |f|$ for all n. Since

$$\int_a^b \left[\sum_{n=1}^k f_n(x) \right] dx = 0 \qquad \text{for all } k$$

by the Beppo Levi Theorem 5.3, $\sum_{n=1}^\infty f_n$ converges almost everywhere on $[a, b]$. Hence $f = 0$ almost everywhere. $\square$

The monotone convergence theorem has a great similarity to our definition of the integral for the class L^+; it is clear that our definition (in particular, the Second Fundamental Lemma) was inspired by this theorem.

We close this section with the following proposition about the integrability of a function defined on $[a, b]$; the proof is left to the reader as an exercise.

5.9. Proposition. *A function f defined on $[a, b]$ belongs to the class L if and only if there is a series $\sum_{n=1}^\infty \varphi_n$ of step functions on $[a, b]$ such that:*

(a) $\sum_{n=1}^\infty \varphi_n = f$ *almost everywhere on* $[a, b]$; *and*
(b) $\sum_{n=1}^\infty \int_a^b |\varphi_n(x)| \, dx$ *converges.*

In this case,

$$\int_a^b f(x) \, dx = \sum_{n=1}^\infty \int_a^b \varphi_n(x) \, dx.$$

The above proposition is similar to the Beppo Levi Theorem 5.7. This can be used as a definition of the class L without introducing the intermediate class L^+ (see Exercise 5D).

EXERCISES 5

A. Let (f_n) be a monotone sequence of functions in the class L which converges to a function in L. Show that

$$\int_a^b [\lim f_n(x)]\, dx = \lim \int_a^b f_n(x)\, dx.$$

B. Let $\sum_{n=1}^\infty f_n$ be a series of functions in L such that $f_n \geq 0$. If $\sum_{n=1}^\infty f_n \in L$, show that

$$\int_a^b \left[\sum_{n=1}^\infty f_n(x) \right] dx = \sum_{n=1}^\infty \int_a^b f_n(x)\, dx.$$

C. Let $f \in L$ and $0 \leq f < 1$. Show that

$$\lim \int_a^b f^n(x)\, dx = 0.$$

D. Define the class L by Proposition 5.9 and derive all the properties discussed in §4 and §5.

E. Let $f \in L$. Show that for every Riemann integrable function g, $fg \in L$. In particular, show that, for every real number α, $f(x) \sin \alpha x$, $f(x) \cos \alpha x \in L$.

F. Show that Theorem 5.3 is false if we do not require the condition $f_n \geq 0$.

G. Show that Theorem 5.7 is false if we replace the condition

$$\sum_{n=1}^\infty \int_a^b |f_n(x)|\, dx < \infty,$$

by

$$-\infty < \sum_{n=1}^\infty \int_a^b f_n(x)\, dx < \infty.$$

H. Let (f_n) be a monotone decreasing sequence of functions in the class L such that $f_n \geq 0$. If $\lim \int_a^b f_n(x)\, dx = 0$, show that $\lim f_n = 0$ almost everywhere.

§6. The Lebesgue Theorem—Dominated Convergence Theorem

As a consequence of the Beppo Levi theorem, we have learned that the class L is closed under the limit passage through monotone increasing sequences of functions whose integrals have a common bound. In this section we shall examine the class L by using sequences of a more general type than those we have considered in the Monotone Convergence Theorem. The requirement in the Monotone Convergence Theorem that the sequence (f_n) be monotone increasing is sometimes very inconvenient; for this purpose one may consider arbitrary nonmonotone sequences and their limit passages. However, the following examples suggest that, in order to insure term-by-term integrability in arbitrary limit passage, it is necessary to take certain precautions.

6.1. Examples. (1) Let $f_n(x) = nx^n$ for $0 \leq x \leq 1$. Then $\lim f_n = 0$ almost everywhere, and $\lim \int_0^1 f_n(x)\, dx = 1$. Therefore,

$$\lim \int_0^1 f_n(x)\, dx \neq \int_0^1 [\lim f_n(x)]\, dx.$$

(2) Let $f_n(x) = n^2 x^n$ for $0 \leq x \leq 1$. Then $\lim f_n = 0$ almost everywhere, and $\lim \int_0^1 f_n(x)\, dx = \infty$. Therefore,

$$\lim \int_0^1 f_n(x)\, dx \neq \int_0^1 [\lim f_n(x)]\, dx$$

(3) Let

$$f_n(x) = \begin{cases} n \sin nx & \text{for } 0 \leq x \leq \pi/n, \\ 0 & \text{for } \pi/n \leq x \leq \pi. \end{cases}$$

The $\lim f_n = 0$ for every x in $[0, \pi]$, and $\lim \int_0^\pi f_n(x)\, dx = 2$. Therefore,

$$\lim \int_0^\pi f_n(x)\, dx \neq \int_0^\pi [\lim f_n(x)]\, dx. \qquad \square$$

In the above examples we notice that none of the sequences are monotone and that none of them are bounded almost everywhere. Therefore it is quite natural to consider the case in which sequences are bounded almost everywhere in order to insure term-by-term integrability. More generally, we have the following important theorem, discovered by Lebesgue (1904) and considered as the *fundamental theorem of the Lebesgue integral*. This theorem again asserts that the class L is closed under the limit procedure we are considering now.

6.2. The Lebesgue Theorem. *If a sequence (f_n) of functions in the class L converges almost everywhere to a function f and if there is a function $g \in L$ such that*

$$|f_n(x)| \leq g(x) \quad a.e. \quad \text{for all } n,$$

then $f \in L$ and

$$\int_a^b f(x)\, dx = \lim \int_a^b f_n(x)\, dx.$$

We use the Monotone Convergence Theorem in this proof. For this we shall construct two monotone sequences from (f_n) which converge almost everywhere to f. The construction of such sequences is given in the following lemma:

6.3. Lemma. *Let (f_n) be a sequence of functions in the class L which converges almost everywhere to a function f. For each n let*

$$g_n = \sup\{f_n, f_{n+1}, \ldots\}$$

and

$$h_n = \inf\{f_n, f_{n+1}, \ldots\}.$$

Suppose that there exists a function $g \in L$ such that

$$|f_n(x)| \leq g(x) \quad a.e. \quad \text{for all n.}$$

Then $g_n, h_n \in L$ and

$$\lim g_n = f = \lim h_n$$

almost everywhere. Consequently, $f \in L$.

Proof of Lemma. For any two functions p and q in L, both $\max\{p, q\}$ and $\min\{p, q\}$ belong to L by Proposition 4.2. Hence, by induction, we have

$$g_{nk} = \max\{f_n, f_{n+1}, \ldots, f_{n+k}\} \in L,$$
$$h_{nk} = \min\{f_n, f_{n+1}, \ldots, f_{n+k}\} \in L.$$

Since $g_{nk} \leq g$ and $-g \leq h_{nk}$, we have

$$\int_a^b g_{nk}(x)\, dx \leq \int_a^b g(x)\, dx \quad \text{and} \quad -\int_a^b g(x)\, dx \leq \int_a^b h_{nk}(x)\, dx$$

for all k. Therefore, by the Monotone Convergence Theorem 5.5 and Corollary 5.6, $\lim_{k \to \infty} g_{nk} = g_n$ and $\lim_{k \to \infty} h_{nk} = h_n$ belong to L.

It is clear that (g_n) is monotone decreasing, (h_n) is monotone increasing, and $\lim g_n = f = \lim h_n$ almost everywhere. Consequently $f \in L$. □

Proof of Theorem. Since both (g_n) and (h_n) are monotone and converging to f almost everywhere, we have

$$\int_a^b f(x)\, dx = \lim \int_a^b g_n(x)\, dx = \lim \int_a^b h_n(x)\, dx$$

by the Monotone Convergence Theorem. On the other hand, $h_n(x) \leq f_n(x) \leq g_n(x)$, and hence

$$\int_a^b h_n(x)\, dx \leq \int_a^b f_n(x)\, dx \leq \int_a^b g_n(x)\, dx.$$

It follows that

$$\lim \int_a^b f_n(x)\, dx = \int_a^b f(x)\, dx.$$

In our proof of the preceding theorem we heavily used the monotone convergence theorem. Historically, of course, Lebesgue proved this theorem starting with his definition, which we still have to show is equivalent to ours.

But one should notice that the Monotone Convergence Theorem is a special case of the preceding theorem. Thus we conclude that the Lebesgue theorem, the Beppo Levi theorem, and the Monotone Convergence Theorem are essentially equivalent. Lebesgue's theorem is usually called the *Lebesgue Dominated Convergence Theorem*. This is one of the triumphs of Lebesgue's theory of integration.

Research on the term-by-term integration of nonuniform converging sequences and series was carried out by eminent mathematicians such as Dini, du Bois–Reymond, Weierstrass, Kronecker, Osgood, and Arzelà since the 1870s. The importance of such research was stressed in the theory of trigonometric series.

The Lebesgue Dominated Convergence Theorem has the following consequence:

6.4. Corollary. *If a sequence (f_n) of functions in the class L converges almost everywhere to a function f and if for some $A > 0, |f_n(x)| < A$ a.e. for all n, then $f \in L$ and $\int_a^b f(x) \, dx = \lim \int_a^b f_n(x) \, dx$.*

We have seen that the preceding corollary is not valid for the Riemann integral (see Example 6.1, Chapter I). The following corollary is stated for the Riemann integral, and follows immediately from Corollary 6.4 and Proposition 3.6.

6.5. Corollary. *If a sequence (f_n) of Riemann integrable functions on $[a, b]$ converges almost everywhere to a Riemann integrable function f and satisfies $|f_n(x)| < A$ almost everywhere for all n, then*

$$\int_a^b f(x) \, dx = \lim \int_a^b f_n(x) \, dx.$$

This result was first proved by the Italian mathematician Cesare Arzelà in 1885 and remained almost unobserved until it was rediscovered independently, in 1897, by W.F. Osgood, in his "Nonuniform convergence and the integration of series term by term." In this memoir all functions are supposed to be continuous. Arzelà's result is of the form of Corollary 6.5, which is more general than that of Osgood; it appeared in "Sulla integrazione per serie" (1885). Of course, their proofs are independent of the Lebesgue theory.

A similar result is the theorem due to P. Fatou, from his "Séries trigonométriques et séries de Taylor" (1906). In this work, the hypothesis $|f_n(x)| \leq g(x)$ in the Lebesgue theorem is replaced by certain other conditions, and we obtain the integrability of the limit function; however, with respect to term-by-term integration it gives only an approximation rather than a precise value.

6.6. Theorem (Fatou's Lemma, 1906). *If (f_n) is a sequence of nonnegative functions in the class L which converges almost everywhere to a function f, and* *if furthermore*

$$\int_a^b f_n(x) \, dx \le A \qquad \text{for all } n,$$

then $f \in L$ *and*

$$\int_a^b f(x) \, dx \le A.$$

Proof. The proof of this lemma is actually the same as that of Lemma 6.3. In fact, consider the functions

$$h_n = \inf\{f_n, f_{n+1}, \ldots\}.$$

Then (h_n) is a monotone increasing sequence which converges almost everywhere to f. Furthermore we have

$$h_n \le f_n \quad \text{and} \quad \int_a^b h_n(x) \, dx \le \int_a^b f_n(x) \, dx \le A$$

and hence, by the Monotone Convergence Theorem, we deduce that $f(x) = \lim h_n \in L$ and $\int_a^b f(x) \, dx \le A$. □

An equivalent formulation of Fatou's lemma needs the concepts of the *limit superior* and *limit inferior* of a sequence (x_n) of real numbers. The limit superior and limit inferior are defined by the following equations:

$$\limsup x_n = \inf\left\{\sup_{n \ge k} x_n : k \in \mathbb{N}\right\},$$

$$\liminf x_n = \sup\left\{\inf_{n \ge k} x_n : k \in \mathbb{N}\right\},$$

where $\sup_{n \ge k} x_n = \sup\{x_n : n \ge k\}$ and $\inf_{n \ge k} x_n = \inf\{x_n : n \ge k\}$. Since every bounded nonempty set of real numbers has a supremum and an infimum, it follows that every bounded sequence of real numbers has both a limit superior and a limit inferior. Conversely, if a sequence has both a limit superior and a limit inferior, then the sequence is bounded. In fact, we have the following relations:

$$\inf\{x_1, x_2, \ldots\} \le \liminf x_n \le \limsup x_n \le \sup\{x_1, x_2, \ldots\}.$$

Now let us state the following theorem from elementary analysis in order to motivate readers to use these limit concepts. The proof is left to the reader.

6.7. Theorem. *For an arbitrary sequence (x_n) of real numbers the following conditions are equivalent:*

(a) *(x_n) converges.*
(b) *Cauchy condition: For every $\varepsilon > 0$, there exists a natural number N such that $|x_n - x_m| < \varepsilon$ holds for every $m, n \ge N$.*
(c) *$-\infty < \liminf x_n = \limsup x_n < \infty$.*

6.8. Theorem (Fatou's Lemma). *If (f_n) is a sequence of nonnegative functions in the class L which converges almost everywhere to a function f, and if furthermore*

$$\int_a^b f_n(x)\, dx \leq A \qquad \text{for all } n,$$

then $f \in L$ and

$$\int_a^b f(x)\, dx \leq \lim\inf \int_a^b f_n(x)\, dx.$$

The proof is almost the same as that of Fatou's Lemma 6.6. It is left to the reader. The sign in the conclusion of Theorem 6.8 may well be strict inequality [see Examples 6.1, (1) to (3)]. Unless the f_n are nonnegative, Fatou's lemma may not hold, even in the presence of uniform convergence.

6.9. Example. Let $f_n(x) = -n$ if $1/n \leq x \leq 2/n$ and 0 otherwise. Then $f = \lim f_n = 0$ almost everywhere and

$$\lim\inf \int_0^1 f_n(x)\, dx = -1 < 0 = \int_0^1 f(x)\, dx.$$

EXERCISES 6

A. Show that there is no sequence of functions on $[0, 2\pi]$ of the type

$$f_n(x) = a_n \sin nx + b_n \cos nx.$$

which converges to the function 1 almost everywhere on $[0, 2\pi]$, and where $|a_n| + |b_n| \leq 10$.

B. Let $f \in L$. Define $f_h(x) = f(x + h)$, $h \in \mathbb{R}$. Show that

$$\lim_{h \to 0} \int_a^b |f_h(x) - f(x)|\, dx = 0.$$

C. Let (f_n) be a sequence of functions in the class L such that

$$\sum_{n=1}^{\infty} |f_n(x)| < \infty \qquad \text{for almost all } x \in [a, b].$$

Show that the series $\sum_{n=1}^{\infty} f_n$ converges almost everywhere to a function in the class L and

$$\int_a^b \left[\sum_{n=1}^{\infty} f_n(x) \right] dx = \sum_{n=1}^{\infty} \int_a^b f_n(x)\, dx.$$

D. Let (f_n) be a sequence of functions in the class L and let $f \in L$ be such that

$$\lim \int_a^b |f_n(x) - f(x)|\, dx = 0.$$

Show that if $f_n \to g$ almost everywhere, then $f = g$ almost everywhere.

E. Give a proof of Theorem 6.2 using Fatou's Lemma 6.8.

F. Prove Theorem 6.7.

G. Prove Theorem 6.8.

H. Prove that if (f_n) is a sequence in L which converges almost everywhere to a function f, and $|f| \leq g$ for some $g \in L$, then $f \in L$.

§7. The Space L^1

In this section we will study the space L of Lebesgue integrable functions defined on $[a, b]$ and answer some questions raised in §6, Chapter I. It will be convenient to introduce the concept of normed spaces at this stage and state the main theorem of this section in terms of normed spaces.

Let E be a vector space over the field $\mathbb{R}$. We introduce into such a space (when possible) a norm function.

7.1. Definition. A real-valued function p defined on a vector space E is said to be a *norm* if it satisfies the following properties:

(a) $p(x) \geq 0$ for all $x \in E$;
(b) $p(x) = 0$ if and only if $x = 0$, the zero vector;
(c) $p(\alpha x) = |\alpha| p(x)$ for all $x \in E$ and $\alpha \in \mathbb{R}$; and
(d) $p(x + y) \leq p(x) + p(y)$ for all $x, y \in E$.

If p is a norm on E, it is customary to denote $p(x)$ by $\|x\|$. The notation $\|x\|$ will henceforth be preferred for the norm of the element $x \in E$.

Property (d) is known as *the triangle inequality*. To understand properties (a) to (d) of a norm, replace $p(x)$ with $\|x\|$ and compare these properties with the properties of the absolute value $|x|$ of real numbers x. You will notice that a norm is a simple generalization of the absolute value.

A vector space with a norm will be called a *normed space*. Every normed space E becomes a metric space if we define a distance

$$d(x, y) = \|x - y\|$$

for all $x, y \in E$. The fact that d is a metric follows at once from properties (a) to (d). Thus everything said about metric spaces in §7, Chapter Zero carries over to the case of normed spaces. Examples 7.2 and 7.3, Chapter Zero, are normed spaces if we define

$$\|x\| = d(x, 0),$$

$$\|f\| = d(f, 0),$$

respectively.

One of the pioneering workers in this subject was the Polish mathematician Stefan Banach (1892–1945), author of the classic *Théorie des Opérations Linéaires* (1932). In honor of Banach, we have the following definition:

7.2. Definition. A normed space E is called a *Banach space* if E is complete (see Definition 7.8, Chapter Zero); that is, every Cauchy sequence in E converges with respect to the metric

$$d(x, y) = \|x - y\|.$$

An obvious example of a Banach space is the n-dimensional Euclidean space $\mathbb{R}^n$. We will present now the most important example of a Banach space.

It was shown that the space L of Lebesgue integrable functions on $[a, b]$ is a vector space over the field $\mathbb{R}$; i.e., if $f, g \in L$ and $\alpha, \beta \in \mathbb{R}$, then $\alpha f + \beta g \in L$. For a function $f \in L$, we define

$$\|f\| = \int_a^b |f(x)| \, dx.$$

Then:

(a) $\|f\| \geq 0$;
(b) $\|f\| = 0$ if and only if $f = 0$ almost everywhere;
(c) If $\alpha \in \mathbb{R}$, then $\|\alpha f\| = |\alpha| \|f\|$; and
(d) $\|f + g\| \leq \|f\| + \|g\|$.

Unfortunately, we can only conclude that $\|f\| = 0$ if and only if $f = 0$ almost everywhere. Therefore the function $f \to \|f\|$ is not a norm on L. We shall, however, consider two integrable functions to be *equivalent* if they are equal almost everywhere; then if we do not distinguish between equivalent functions, the space L becomes a normed space. We denote this normed space by L^1, or more precisely $L^1[a, b]$, and call the norm $\|\cdot\|$ the L^1 norm. To be pedantic we should say that the elements of L^1 are not functions but rather equivalence classes of functions. We shall avoid such unnecessary pedantry in the future by speaking simply of integrable functions rather than the equivalence classes of integrable functions.

We have the following main theorem:

7.3. Theorem (Riesz, 1910). *The space L^1 is a Banach space.*

Proof. Since L^1 is a normed space, it remains to show that every Cauchy sequence in L^1 converges to a function in L^1. Let (f_n) be a Cauchy sequence. Then, there is a natural number n_1 such that for all $n \geq n_1$, we have

$$\|f_n - f_{n_1}\| < \tfrac{1}{2}.$$

By induction, after finding $n_{k-1} > n_{k-2}$, we find $n_k > n_{k-1}$ such that for all $n > n_k$ we have

$$\|f_n - f_{n_k}\| < \frac{1}{2^k}.$$

Then (f_{n_k}) is a subsequence of (f_n) which satisfies

$$\| f_{n_{k+1}} - f_{n_k} \| < \frac{1}{2^k} \qquad \text{for all } k.$$

That is,

$$\int_a^b | f_{n_{k+1}}(x) - f_{n_k}(x) | \, dx < \frac{1}{2^k} \qquad \text{for all } k.$$

This implies that

$$\int_a^b | f_{n_1}(x) | \, dx + \sum_{k=1}^{\infty} \int_a^b | f_{n_{k+1}}(x) - f_{n_k}(x) | \, dx < \infty.$$

But then, by the Beppo Levi Theorem 5.7,

$$f_{n_1} + \sum_{k=1}^{\infty} [f_{n_{k+1}}(x) - f_{n_k}(x)]$$

converges almost everywhere to an integrable function f, and

$$\int_a^b f(x) \, dx = \int_a^b f_{n_1}(x) \, dx + \sum_{k=1}^{\infty} \int_a^b [f_{n_{k+1}}(x) - f_{n_k}(x)] \, dx.$$

We now show that $\| f_{n_k} - f \| \to 0$ as $k \to \infty$. We first notice that

$$f(x) - f_{n_p}(x) = \sum_{k=p}^{\infty} [f_{n_{k+1}}(x) - f_{n_k}(x)].$$

It follows from this that

$$\| f - f_{n_p} \| \leq \sum_{k=p}^{\infty} \| f_{n_{k+1}} - f_{n_k} \| < \sum_{k=p}^{\infty} \frac{1}{2^k} = \frac{1}{2^{p-1}}$$

showing that $\| f - f_{n_p} \| \to 0$ as $p \to \infty$. Finally, we prove $\| f - f_n \| \to 0$ as $n \to \infty$. This is easy, for

$$\| f_n - f \| \leq \| f_n - f_{n_k} \| + \| f_{n_k} - f \|,$$

where $\| f_n - f_{n_k} \| \to 0$ as $n \to \infty$ and $k \to \infty$, and hence $\| f_n - f \| \to 0$ as $n \to \infty$. $\qquad \square$

Historically, in 1907 F. Riesz and E. Fischer independently published a theorem similar to the preceding one for a space of square integrable functions, that is, $f^2 \in L$, in their study of Fourier series (see §6, Chapter VI). Later, in 1910, Riesz generalized the result for the space L^p of functions f such that $|f|^p \in L$, $1 \leq p < \infty$. We will study such spaces in Chapter VI; we only remark here that the preceding Riesz theorem is extremely important because it secured a permanent place for the new theory of integration in the development of functional analysis.

We will now show that the Lebesgue space L^1 is the completion of the space $C[a, b]$ of continuous functions on $[a, b]$ with respect to the L^1 norm. First consider the space $C[a, b]$ with the L^1 norm. Since $f \in C[a, b]$ and $\|f\| = \int_a^b |f(x)|\, dx = 0$ imply that f is identically 0, the L^1 norm makes $C[a, b]$ a normed space. We have shown that this space is not complete (see p. 47). Since every continuous function is Lebesgue integrable, we can easily identify $C[a, b]$ as a subspace of L^1. To show that L^1 is the completion of the space $C[a, b]$, we must prove that every Lebesgue integrable function is a limit of a sequence of continuous funcitons with respect to the L^1 norm. This is in fact equivalent to the following proposition:

7.4. Proposition. *Let $f \in L$ and $\varepsilon > 0$. Then there exists a function $g \in C[a, b]$ such that $\|f - g\| < \varepsilon$; that is, $C[a, b]$ is dense in L.*

Proof. We prove this proposition in three steps.

Step 1. Let φ be a step function on $[a, b]$ and $\varepsilon > 0$. Then there exists a continuous function g on $[a, b]$ such that

$$\|\varphi - g\| < \varepsilon.$$

Figure 2.4 may guide the reader to an easy proof of the above statement. The details are left to the reader.

Step 2. Let $f \in L^+$ and $\varepsilon > 0$. Then there exists a step function φ on $[a, b]$ such that

$$\|f - \varphi\| < \varepsilon.$$

In fact, choose a monotone increasing sequence (φ_n) of step functions converging almost everywhere to f. Then $(f - \varphi_n)$ is monotonic decreasing to 0 almost everywhere, and hence, by Corollary 5.6 to the monotone convergence theorem, we have

$$\lim \int_a^b |f(x) - \varphi_n(x)|\, dx = \lim \int_a^b [f(x) - \varphi_n(x)]\, dx = 0$$

which proves the statement of Step 2.

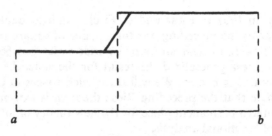

Figure 2.4

Step 3. Now we prove the proposition. Let $f \in L$ and write $f = f_1 - f_2$, where $f_1, f_2 \in L^+$. For $\varepsilon > 0$ find step functions φ_1 and φ_2 such that $\|f_j - \varphi_j\| < \varepsilon/3, j = 1, 2$, by Step 2. Then, find a continuous function g, by Step 1, such that $\|g - (\varphi_1 - \varphi_2)\| < \varepsilon/3$. Thus we have

$$\begin{aligned}
\|f - g\| &= \|(f_1 - f_2) - g\| \\
&\leq \|(f_1 - f_2) - (\varphi_1 - \varphi_2)\| + \|(\varphi_1 - \varphi_2) - g\| \\
&\leq \|f_1 - \varphi_1\| + \|f_2 - \varphi_2\| + \|(\varphi_1 - \varphi_2) - g\| \\
&< \varepsilon/3 + \varepsilon/3 + \varepsilon/3 = \varepsilon. \qquad \square
\end{aligned}$$

The preceding result is also true if we replace continuous functions by indefinitely differentiable functions of compact support (see Exercise 7D).

We close this section by giving an example of a Lebesgue integrable function which is not equivalent to any Riemann integrable function. The reader may recall that the Dirichlet function which takes 1 on rationals and 0 on irrationals is not Riemann integrable, but it is equivalent to the constant function 0 (zero), which is Riemann integrable.

7.5. Proposition. *There is a Lebesgue integrable function on* $[0, 1]$ *which is not equivalent to any Riemann integrable function.*

Proof. Let $\{r_1, r_2, \ldots\}$ be the set of all rationals in $(0, 1)$. For each n, let I_n be an open interval in $(0, 1)$ such that $r_n \in I_n$ and $|I_n| = 1/2^{n+1}$. Let U be the union of $I_1, I_2, \ldots$. Then U is an open and dense set in $[0, 1]$ which is not of measure zero. Furthermore, its complement $[0, 1] \backslash U$ is not of measure zero. Let f be the characteristic function of U. Then f is Lebesgue integrable on $[0, 1]$.

We now claim that there is no Riemann integrable function equivalent to f. Suppose that g is a function on $[0, 1]$ such that $f = g$ almost everywhere. We want to show that g is not Riemann integrable on $[0, 1]$. Let $D = \{x \in [0, 1]: f(x) \neq g(x)\}$. Then D is of measure zero. Since U is dense in $[0, 1]$, so is $U \backslash D$ (why?). Then

$$g(x) = \begin{cases} 1 & \text{if } x \in U \backslash D, \\ 0 & \text{if } x \in ([0, 1] \backslash U) \backslash D. \end{cases}$$

Therefore, g is not continuous at any point of $([0, 1] \backslash U) \backslash D$. This shows that the set of discontinuities of g is not of measure zero; that is, g is not Riemann integrable. $\qquad \square$

7.6. Proposition. *A function f on $[a, b]$ is equivalent to a Riemann integrable function if and only if $f \in L^+$ and $-f \in L^+$.*

Proposition 7.6 completely characterizes the equivalence class of a Riemann integrable function. The necessary condition was in Corollary 3.7. The remaining part is an easy exercise (see Exercise 3E).

EXERCISES 7

A. Let $f_n \in L^1$ be such that

$$\sum_{n=1}^{\infty} \int_a^b |f_n(x)| \, dx < \infty.$$

Show that there is a function $f \in L^1$ such that

$$\sum_{n=1}^{\infty} \int_a^b f_n(x) \, dx = \int_a^b f(x) \, dx.$$

B. Let $B = \{f \in L^1[0, 1]: \|f\| \leq 1\}$. Give an example of a sequence in B which has no convergent subsequences with respect to the L^1 norm. (This proves that B is not compact, and hence the L^1 space is of infinite dimension. Such a study is not our intention. The interested reader is referred to any book on functional analysis.)

C. Let $f = \chi_{[a, b]}$. Show that for $\varepsilon > 0$ there is a function $g: \mathbb{R} \to \mathbb{R}$ such that:
   (1) g is indefinitely differentiable;
   (2) g has compact support; i.e., g vanishes on the complement of a compact set; and
   (3) $\int_a^b |f(x) - g(x)| \, dx < \varepsilon$.
   Hint: See §2, Chapter VI, if you make no progress.

D. Let f be integrable on $[a, b]$, and let $\varepsilon > 0$. Show that there is an indefinitely differentiable function $g: \mathbb{R} \to \mathbb{R}$ such that

$$\int_a^b |f(x) - g(x)| \, d \; < \varepsilon$$

Appendix

Henri Lebesgue (1875–1941). Henri Lebesgue was born in Beauvais, fifty miles from Paris, on June 28, 1875. His father was a typesetter and his mother was an elementary school teacher. He was so promising that the town sponsored his education in spite of the early death of his father, and thus indirectly sponsored one of the best theses ever written.

Lebesgue received his early training at the École Normale Supérieure. His first university posts were at Rennes and later (1906) at Poitiers as professor in the Faculté des Sciences. He then went to Paris in 1910 and afterward was professor at the Collège de France. Lebesgue was elected to the Académie des Sciences in 1922, made an honorary member of the London Mathematical Society two years later, and named foreign member of the Royal Society in 1930.

Lebesgue's early work was published for the most part in *Comptes Rendus de l'Académie des Sciences* between 1898 and 1901. In several notes, Lebesgue considered primarily problems of plane surfaces and integrations. The results of this productive period may be considered as preliminaries to Lebesgue's thesis on measure and integration; in 1901, for example, Lebesgue published his procedure of partitioning the range of a function $f(x)$ rather than its domain—a process uncommon to methods of Riemann integration—in defining the Lebesgue integral of $f(x)$ (Lebesgue, 1901). Then, in 1902, Lebesgue's collected study concerning integration, measure, curve length, primitive function, etc., a paper entitled "Intégrale, longeur, aire," appeared in the *Annali di Mathematica*. With this basis, Lebesgue lectured on his theory of integration at the Collège de France, 1902–1903, after which time his lecture notes were compiled into a Borel tract, *Leçons sur l'Intégration et la Recherche des Fonctions Primitives* (1904). This major work, which Lebesgue accumulated in the span of five years, served without a doubt to revolutionize modern concepts of integration and analysis of functions, including applications to mathematics and more widely to various outside fields of science.

After publication and presentation of his formidable thesis, Lebesgue continued, by means of university lectures and his personal research, to illustrate the scope and applicability of his integration and measure theory. He was led immediately into the analysis of convergence of trigonometric series; he later dealt in detail with measurable sets, Baire classes, and the foundations of his definition of integral and relations therein to his concept of measure. Lebesgue's work was centered almost exclusively on functions of real variables; from 1907 to 1912 he considered, but not to an exhaustive extent, questions of topology and complex variables in research on potential theory. In the later years of his life, his major endeavors having been enlarged upon quite competently both by himself and by his colleagues, Lebesgue turned to less taxing mathematical problems of wider interest, having to do primarily with somewhat elementary geometry. In 1928 he did revise and expand comprehensively his *Lecons sur l'Intégration*, although the foundations of this excellent work had already been laid.

Lebesgue may have been somewhat restricted, in terms of being a universal mathematician, in his having spent most of his efforts on the analysis of real functions; because of his innovative and extensive techniques in that field, however, Lebesgue truly remains a foremost contributor to modern mathematics.

For a further study of Lebesgue's life and work we refer the interested reader to the "Biographical Sketch of Henri Lebesgue," by Kenneth O. May, contained in Lebesgue (1966).

Frigyes Riesz (1880–1956). Frigyes Riesz was born in Györ, Hungary, on January 22, 1880. He studied at the Polytechnic in Zürich (1897–1899), in Budapest (1899–1901), and in Göttingen and Paris (1903–1904). He became a high-school teacher after his Ph.D degree. His doctoral thesis was on projective geometry. In 1907 Riesz gained widespread recognition through the Riesz–Fischer theorem. As early as 1908 Riesz formulated the general axioms of topological space; to him is due, in part, the concept of T_1 spaces. In 1910, Riesz introduced and studied the spaces L^p for $1 \leq p \leq \infty$. This study secured for the Lebesgue theory of integration a permanent place in functional analysis. He became a professor at the University of Kolzvál (now Cluj, Rumania) in 1911. After Kolzvál became a part of Rumania, he moved to Szeged and founded János Bolyai Mathematical Institute in 1920. In 1944 he became President (Rector) of Szeged University. With Alfred Haar, Riesz founded *Acta Universitatis Szegediensis*, a first Hungarian journal strictly devoted to mathematics. He became a member of the Budapest Academy in 1936. In 1945, he was appointed Professor of Mathematics at the University of Budapest. He died on February 28, 1956. His younger brother, Marcel Riesz, was also a distinguished mathematician. He made notable contributions in numerous areas in functional analysis.

CHAPTER III
Lebesgue Measure

The Lebesgue theory originally was based on an improvement and general-ization of the work of Emil Borel, *Leçons sur la Théorie des Fonctions* (1895). Borel had already presented a theory of measure for the class of sets now known as Borel sets.

The classical method of developing the Lebesgue theory proceeds in the following way: First, with the aid of the simple concept of the length of an interval, we construct a real-valued function called an *outer measure*. This function, which we shall denote by m^*, has for its domain the power set of $[a, b]$. The function m^* is not countably additive. Then we attempt to diminish the domain of m^* in such a way that m^* actually becomes countably additive on the reduced domain (see, §10). A set in the reduced domain is called measurable. Next follows the definition of measurable functions, and finally the definition of summable (or integrable) functions and their integrals.

In this chapter we will study Lebesgue measure as a consequence of the theory of integration with which we have become acquainted in Chapter II.

§1. Measurable Functions

In Proposition 4.7, Chapter II, we have shown that if f is integrable, then there exists a sequence (φ_n) of step functions defined on $[a, b]$ such that $f = \lim \varphi_n$ almost everywhere on $[a, b]$; i.e., every integrable function is a limit of a sequence of step functions. It is easy to see that the converse is not true. For example, consider the function

$$f(x) = \begin{cases} 1/x & \text{if } x \in (0, 1], \\ 0 & \text{if } x = 0. \end{cases}$$

Then f is not integrable; however, it is representable almost everywhere as a limit of a sequence of step functions (see Exercise 1A).

We have the following definition:

1.1. Definition. A function $f\colon [a, b] \to \mathbb{R}^*$ is said to be *measurable* if it can be represented almost everywhere as a limit of a sequence $\{\varphi_n\}$ of step functions which converges almost everywhere on $[a, b]$.

We have a proposition that follows immediately from the definition.

1.2. Proposition. *Every integrable function on* $[a, b]$ *is measurable. In particular, every continuous function on* $[a, b]$ *is measurable.*

1.3. Proposition. *Let* f *and* g *be measurable functions and let* c *be a real number. Then the functions*

$$cf, \quad f^2, \quad f + g, \quad fg, \quad |f|, \quad \max\{f, g\}, \quad \min\{f, g\},$$

are also measurable. Furthermore, if g *is nonzero almost everywhere, then* f/g *is measurable.*

Proof. It follows immediately from the definition that $cf, f^2, f + g, fg,$ and $|f|$ are measurable. Since

$$\max\{f, g\} = \tfrac{1}{2}(f + g + |f - g|),$$
$$\min\{f, g\} = \tfrac{1}{2}(f + g - |f - g|),$$

$\max\{f, g\}$ and $\min\{f, g\}$ are measurable.

If g is the limit almost everywhere of the sequence (φ_n) of step functions then $1/g$ will be the limit almost everywhere of the sequence (ψ_n) of step functions which is defined as follows:

$$\psi_n(x) = \begin{cases} 0 & \text{if } \varphi_n(x) = 0, \\ 1/\varphi_n(x) & \text{if } \varphi_n(x) \neq 0. \end{cases}$$

Therefore, $1/g$ is measurable, and hence f/g is measurable. $\quad\square$

1.4. Corollary. f *is measurable if and only if* f^+ *and* f^- *are measurable.*

Proof. Since $f^+ = \max\{f, 0\}$ and $f^- = \max\{-f, 0\}$, if f is measurable, then both f^+ and f^- are measurable by the preceding proposition. The converse is immediate from the identity $f = f^+ - f^-$. $\quad\square$

1.5. Proposition. *Let* f *be a measurable function. Suppose that there exists a function* g *in the class* L *such that*

$$|f(x)| \leq g(x)$$

for almost all x in [a, b]. Then f ∈ L. In particular, every bounded measurable function on [a, b] is integrable.

Proof. Let (φ_n) be a sequence of step functions that defines the measurable function f. Consider

$$\psi_n = \max\{-g, \min\{\varphi_n, g\}\}.$$

This is just another way of saying that ψ_n is obtained by truncating above by g and below by $-g$. Of course, $\psi_n \in L$ and $|\psi_n(x)| \le g(x)$. Since $f = \max\{-g, \min\{f, g\}\}$, it is easy to see that $f = \lim \psi_n$ almost everywhere on $[a, b]$. Therefore, we conclude that $f \in L$ by the Lebesgue Dominated Convergence Theorem.   □

1.6. Proposition. *Let (f_n) be a sequence of measurable functions which converges to a function f almost everywhere. Then f is also measurable.*

Proof. Let f_n be measurable and $f = \lim f_n$ almost everywhere. Consider a function $h \in L$ which is strictly positive. [For this proof $h = 1$ will do, but we hope to use this same proof for measurable functions defined on the entire real line (see §3, Chapter IV).] Then the sequence of functions

$$g_n = \frac{h f_n}{h + |f_n|} \quad \rightarrow \quad \frac{h f}{h + |f|} = g$$

as $n \to \infty$. The functions g_n are measurable by Proposition 1.3. Furthermore, we have

$$|g_n(x)| \le h(x).$$

Therefore $g_n \in L$ by Proposition 1.5, and hence, by the Lebesgue Dominated Convergence Theorem, $g \in L$. It follows that g is measurable. But then from

$$f = \frac{hg}{h - |g|}$$

f is measurable.   □

The reader will recall that the class L of integrable functions on $[a, b]$ is closed under the operation described in the Lebesgue theorem, but not under that of taking pointwise limits. Proposition 1.6 tells us that the class of measurable functions on $[a, b]$ is closed under the operation of taking limits almost everywhere.

We close this section with the following important property of the Lebesgue integral, which claims the Lebesgue integral operates only on those measurable functions whose absolute function is integrable. Compare this with the Riemann theory.

1.7. Proposition. *Let f be a measurable function on* $[a, b]$. *Then f is integrable on* $[a, b]$ *if and only if* $|f|$ *is integrable on* $[a, b]$.

Proof. If f is integrable, then it is already shown that $|f|$ is integrable. Conversely, if $|f|$ is integrable, then f is integrable by Proposition 1.5.   □

1.8. Corollary. *Let f be integrable on* $[a, b]$ *and let g be a measurable function which is bounded almost everywhere on* $[a, b]$. *Then fg is integrable on* $[a, b]$.

Proof. If $|g(x)| \leq A$ almost everywhere, then $|f(x)g(x)| \leq A|f(x)|$ almost everywhere on $[a, b]$. Hence the result follows from Proposition 1.7.   □

EXERCISES 1

A. Show that the function f, defined by $f(x) = 1/x$ if $x \in (0, 1]$ and $f(0) = 0$, is not integrable on $[0, 1]$.

B. Show that a function $f: [a, b] \to \mathbb{R}$ is measurable if and only if for every step function φ, $\max\{-\varphi, \min\{\varphi, f\}\}$ is integrable on $[a, b]$.

C. Suppose that f is measurable, and g and h are integrable on $[a, b]$. Then $\max\{h, \min\{f, g\}\}$ is integrable on $[a, b]$.

D. Suppose that f is measurable on $[a, b]$ such that for every $g \in L$, $fg \in L$. Show that f is bounded almost everywhere. (*Hint:* If you make no progress for this problem, try it again after §4.2.)

E. If g and h are integrable on $[a, b]$, if f is measurable, and if $g(x) \leq f(x) \leq h(x)$ almost everywhere, then f is integrable on $[a, b]$.

§2. Lebesgue Measure

We recall the definition of the characteristic function χ_E of an arbitrary set E:

$$\chi_E(x) = \begin{cases} 1 & \text{if } x \in E, \\ 0 & \text{if } x \notin E. \end{cases}$$

2.1. Definition. A set E contained in the closed interval $[a, b]$ is called *measurable* (or *Lebesgue measurable*) if its characteristic function χ_E is measurable. The *measure* $m(E)$ of the measurable set E is defined by

$$m(E) = \int_a^b \chi_E(x)\, dx.$$

Since if $E \subset [a, b]$ and $E \subset [c, d]$ then

$$\int_a^b \chi_E(x)\, dx = \int_c^d \chi_E(x)\, dx,$$

the measure $m(E)$ is independent of the choice of the intervals which contain E.

According to this definition every finite interval is measurable, and its measure is simply the length of the interval; i.e.,

$$m((c, d)) = m([c, d]) = m((c, d]) = m([c, d)) = d - c.$$

We have already introduced the *set of measure zero* in §4, Chapter I. Suppose that E is a bounded set of measure zero. Then $\chi_E = 0$ almost everywhere, and hence χ_E is integrable. Its integral is zero. Therefore E is measurable and its measure is zero. Conversely, if E is a bounded measurable set and its measure is zero, then $\int_a^b \chi_E(x)\, dx = 0$. This implies that $\chi_E = 0$ almost everywhere by Proposition 5.8, Chapter II, and hence E is a set of measure zero. Therefore, the old definition of a set of measure zero is identical with the present definition for a bounded measurable set whose measure is zero.

The measure just defined possesses the following important properties:

2.2. Proposition. *If E, F, and E_n, $n \in \mathbb{N}$, are measurable sets in $[a, b]$, then*

$$[a, b] \setminus E, \quad E \cup F, \quad E \cap F, \quad E \setminus F, \quad \bigcup_{n=1}^{\infty} E_n, \quad \bigcap_{n=1}^{\infty} E_n,$$

are measurable.

Proof. With E, F, and E_n, $n \in \mathbb{N}$, as described, χ_E, χ_F, and χ_{E_n} are in L, so that

$$\chi_{[a,b] \setminus E} = 1 - \chi_E,$$

$$\chi_{E \cup F} = \max\{\chi_E, \chi_F\},$$

$$\chi_{E \cap F} = \min\{\chi_E, \chi_F\},$$

$$\chi_{E \setminus F} = \chi_E - \chi_{E \cap F},$$

are in L. Therefore, $[a, b] \setminus E$, $E \cup F$, $E \cap F$, and $E \setminus F$ are measurable sets.
For the sequence (E_n), let

$$A = \bigcup_{n=1}^{\infty} E_n \quad \text{and} \quad B = \bigcap_{n=1}^{\infty} E_n.$$

Denote $\chi_n = \chi_{E_n}$. Then

$$\chi_A = \lim \max\{\chi_1, \chi_2, \ldots, \chi_n\}$$

and

$$\chi_B = \lim \min\{\chi_1, \chi_2, \ldots, \chi_n\},$$

where both $\max\{\chi_1, \chi_2, \ldots, \chi_n\}$ and $\min\{\chi_1, \chi_2, \ldots, \chi_n\}$ are measurable. Therefore, by Proposition 1.6, both χ_A and χ_B are measurable, which proves that both A and B are measurable. $\square$

2.3. Proposition. *The measure m is countably additive. This means that if (E_n) is a sequence of measurable sets which are mutually disjoint, then*

$$m\left(\bigcup_{n=1}^{\infty} E_n\right) = \sum_{n=1}^{\infty} m(E_n).$$

Proof. For the sequence (E_n) of mutually disjoint measurable sets, let $E = \bigcup_{n=1}^{\infty} E_n$. We have

$$\chi_E = \sum_{n=1}^{\infty} \chi_{E_n} \le 1.$$

Let $f_k = \sum_{n=1}^{k} \chi_{E_n}$. Then (f_k) is a monotone increasing sequence of functions in the class L converging to χ_E. Hence $\chi_E \in L$ by the Monotone Convergence Theorem, so that E is measurable and

$$m(E) = \int_a^b \chi_E(x)\, dx = \lim \int_a^b f_k(x)\, dx$$

$$= \lim \int_a^b \left[\sum_{n=1}^{k} \chi_{E_n}(x)\right] dx$$

$$= \lim \sum_{k=1}^{k} \int_a^b \chi_{E_n}(x)\, dx = \lim \sum_{n=1}^{k} m(E_n)$$

$$= \sum_{n=1}^{\infty} m(E_n). \qquad \square$$

2.4. Proposition. *If (E_n) is a sequence of measurable sets, then*

$$m\left(\bigcup_{n=1}^{\infty} E_n\right) \le \sum_{n=1}^{\infty} m(E_n).$$

Proof. If (E_n) is an arbitrary sequence of measurable sets, then set $F_1 = E_1$, $F_2 = E_2 \cap ([a, b] \setminus E_1)$, and, in general,

$$F_n = E_n \cap \left([a, b] \setminus \bigcup_{j=1}^{n-1} E_j\right).$$

Then (F_n) is a sequence of mutually disjoint measurable sets such that

$$\bigcup_{n=1}^{\infty} F_n = \bigcup_{n=1}^{\infty} E_n.$$

Hence $\bigcup_{n=1}^{\infty} F_n$ is measurable. Since $F_n \subset E_n$, then by Exercise 2A

$$m\left(\bigcup_{n=1}^{\infty} E_n\right) = m\left(\bigcup_{n=1}^{\infty} F_n\right) = \sum_{n=1}^{\infty} m(F_n) \le \sum_{n=1}^{\infty} m(E_n). \qquad \square$$

2.5. Proposition.

(a) *If* (E_n) *is an increasing sequence of measurable sets, that is,* $E_1 \subset E_2 \subset \cdots$, *then*

$$m\left(\bigcup_{n=1}^{\infty} E_n\right) = \lim m(E_n).$$

(b) *If* (F_n) *is a decreasing sequence of measurable sets, that is,* $F_1 \supset F_2 \supset \cdots$, *then*

$$m\left(\bigcap_{n=1}^{\infty} F_n\right) = \lim m(F_n).$$

Proof. (a) Let $A_1 = E_1$ and $A_n = E_n \setminus E_{n-1}$ for $n > 1$. Then (A_n) is a mutually disjoint sequence of measurable sets such that

$$E_n = \bigcup_{j=1}^{n} A_j, \qquad \bigcup_{n=1}^{\infty} E_n = \bigcup_{n=1}^{\infty} A_n.$$

Since the measure m is countably additive, we have

$$m\left(\bigcup_{n=1}^{\infty} E_n\right) = \sum_{n=1}^{\infty} m(A_n) = \lim \sum_{j=1}^{n} m(A_j) = \lim m(E_n).$$

(b) Let $E_n = F_1 \setminus F_n$, so that (E_n) is an increasing sequence of measurable sets. Applying the assertion of part (a), we have

$$m\left(\bigcup_{n=1}^{\infty} E_n\right) = \lim m(E_n) = \lim[m(F_1) - m(F_n)]$$

$$= m(F_1) - \lim m(F_n).$$

On the other hand, $\bigcup_{n=1}^{\infty} E_n = F_1 \setminus \bigcap_{n=1}^{\infty} F_n$, and it follows that

$$m\left(\bigcup_{n=1}^{\infty} E_n\right) = m(F_1) - m\left(\bigcap_{n=1}^{\infty} F_n\right).$$

Combining these two equations, we obtain the desired identity. □

The countable additivity of the measure m (Proposition 2.3) and Proposition 2.5 are equivalent. The proof of Proposition 2.5 indicates the validity of one implication. The proof of the converse, that Proposition 2.5 implies the countable additivity of the measure, is an easy exercise (see Exercise 2G). An immediate proof of Proposition 2.5 without using the countable additivity of the measure m comes from the Lebesgue Dominated Convergence Theorem if we apply this theorem to the sequence (χ_{E_n}).

EXERCISES 2

A. Let A and B be measurable sets in $[a, b]$. Show that $m(A) \leq m(B)$ if $A \subset B$.

B. Let A and B be measurable sets in $[a, b]$. Show that

$$m(A) + m(B) = m(A \cap B) + m(A \cup B).$$

C. Let A and B be measurable sets in $[a, b]$. Denote

$$A \triangle B = (A \setminus B) \cup (B \setminus A).$$

Prove that $m(A \triangle B) = 0$ if and only if $m(A \setminus B) = 0$ and $m(B \setminus A) = 0$.

D. Prove that if $m(A \triangle B) = 0$, then

$$m(A) = m(A \cap B) = m(B).$$

E. Let A, B, and C be measurable sets in $[a, b]$. Show that if $m(A \triangle B) = 0$ and $m(B \triangle C) = 0$, then $m(A \triangle C) = 0$. Hint: $A \setminus C \subset (A \setminus B) \cup (B \setminus C)$.

F. Let $\mathcal{M}$ be the family of all measurable sets in $[a, b]$. Let $A \sim B$ mean that $m(A \triangle B) = 0$ for A and B in $\mathcal{M}$. Prove that $\sim$ is an equivalence relation on $\mathcal{M}$.

G. Prove that Proposition 2.5(a) implies the countable additivity of the measure m.

H. Prove Proposition 2.5(a) without assuming the countable additivity of the measure m.

I. Let $E \subset [a, b]$ be a measurable set such that both E and its complement are dense in $[a, b]$. Show that the characteristic function χ_E is Lebesgue integrable, but is not Riemann integrable.

J. Construct a measurable set E satisfying the preceding exercise.

K. Use the measure m and show that every open interval is uncountable.

L. Show that open sets and closed sets in $[a, b]$ are measurable.

M. Using the Heine–Borel theorem, show that every compact set in $\mathbb{R}$ is measurable.

§3. σ-Algebras and Borel Sets

We have shown that the class of all measurable sets is a σ-algebra; that is, the class is closed under complementation and countable unions (see Proposition 2.2). In this section we will define the notion σ-algebra for a general set X.

3.1. Definition. A collection $\mathcal{A}$ of subsets of X is called an *algebra* (or *Boolean algebra*) if:

(a) $E \cup F$ is in $\mathcal{A}$ whenever E and F are.
(b) CE is in $\mathcal{A}$ whenever E is.
   Since
$$E \setminus F = E \cap CF = C(CE \cup F), \qquad E \cap F = E \setminus CF,$$
   it follows that:
(c) $E \cap F$ is in $\mathcal{A}$ whenever E and F are.

If a collection $\mathcal{A}$ of subsets of X satisfies (b) and (c), then it also satisfies (a). Therefore $\mathcal{A}$ is an algebra.

By induction, it is also easy to see that if $\mathcal{A}$ is an algebra and $E_1, \ldots, E_n$ are sets in $\mathcal{A}$, then $E_1 \cup \cdots \cup E_n$ and $E_1 \cap \cdots \cap E_n$ are in $\mathcal{A}$.

The following proposition is very useful:

3.2. Proposition. *If $\mathcal{C}$ is any collection of subsets of X, then there exists a unique algebra $\mathcal{A}$ containing $\mathcal{C}$ such that, if $\mathcal{B}$ is any algebra containing $\mathcal{C}$, then $\mathcal{A} \subset \mathcal{B}$.*

The unique algebra $\mathcal{A}$ containing $\mathcal{C}$ in the proposition is called the algebra generated by $\mathcal{C}$.

Proof. It is clear that there is at least one algebra containing $\mathcal{C}$. (In fact, the collection of all subsets of X is an algebra.) Let $\mathcal{F}$ be the family of all algebras which contain $\mathcal{C}$. Denote

$$\mathcal{A} = \bigcap \{\mathcal{B} : \mathcal{B} \in \mathcal{F}\}.$$

Then $\mathcal{A}$ is an algebra (why?) containing $\mathcal{C}$. Let $\mathcal{B}$ be an algebra containing $\mathcal{C}$. Then $\mathcal{B} \in \mathcal{F}$ and $\mathcal{B} \supset \mathcal{A}$ by the definition of $\mathcal{A}$. $\square$

3.3. Definition. An algebra $\mathcal{A}$ of sets is called a *σ-algebra* if, whenever (E_n) is a sequence of sets in $\mathcal{A}$, then $\bigcup_{n=1}^{\infty} E_n$ is in $\mathcal{A}$.

From De Morgan's laws it follows that the intersection of a countable collection of sets in $\mathcal{A}$ is again in $\mathcal{A}$.

According to this terminology, Proposition 2.2 states that the *family of all measurable sets in $[a, b]$ is a σ-algebra.*

The statement and proof of Proposition 3.2 remain unaltered if we substitute "σ-algebra" for "algebra," and we have the following proposition:

3.4. Proposition. *If $\mathcal{C}$ is any collection of subsets of X, then there exists a unique σ-algebra $\mathcal{A}$ containing $\mathcal{C}$ such that, if $\mathcal{B}$ is any σ-algebra containing $\mathcal{C}$, then $\mathcal{A} \subset \mathcal{B}$.*

The smallest σ-algebra $\mathcal{A}$ containing $\mathcal{C}$ is called the *σ-algebra generated by $\mathcal{C}$.*

In §5, Chapter Zero, we learned that, although the intersection of any finite collection of open sets is again open, the intersection of a countable collection of open sets may not be open. Therefore, the collection of all open sets is not a σ-algebra. This leads us to the following notion:

3.5. Definition. Let $\mathcal{B}$ be the σ-algebra generated by the family of all intervals in $[a, b]$ (or in $\mathbb{R}$). The elements of $\mathcal{B}$ are called *Borel sets* in $[a, b]$ (or in $\mathbb{R}$).

Therefore, Borel sets can be obtained from the intervals by a countable number of successive operations of taking unions, intersections, or complements. As it happens, there are various ways of defining Borel sets, and the reader of further measure and integration theory must check the definitions of the particular author.

We also notice that every open set (and hence every closed set) is a Borel set.

3.6. Proposition. *Every Borel set in* $[a, b]$ *is measurable. In particular, every open set and every closed set in* $[a, b]$ *is measurable.*

Proof. Since the class of all measurable sets forms a σ-algebra and intervals are measurable, the smallest σ-algebra containing all intervals will be a subclass of the class of all measurable sets. Therefore, every Borel set is measurable. $\square$

3.7. Proposition. *The cardinality of the class of all Borel sets in* $[a, b]$ *is c, the cardinality of* $\mathbb{R}$.

The proof requires transfinite induction. We shall indicate the nature of the argument without giving all the details on which it depends. For general orientation on the subject of ordinal numbers and transfinite induction, we refer the reader to the following sources: Halmos (1960); Wilder (1965). The purpose of giving the following proof is to make our presentation complete. The reader without knowledge of transfinite numbers may skip the proof now, but is strongly urged to study such a subject in the near future.

If X is a nonempty set of cardinality m, the cardinal number of the power set $\mathscr{P}(X)$, the class of all subsets of X, is 2^m. It is a theorem that $m < 2^m$. Also, if $m \geq \aleph_0$, then $m \cdot m = m$. It is not hard to see that $2^{\aleph_0} = c$. The Cantor continuum hypothesis says that $c = \aleph_1$, the first uncountable cardinal number.

Proof. Let E_0 be the collection of all intervals in $[a, b]$. Then the cardinality of E_0 is c. Let E_1 denote the collection of all sets which can be obtained from E_0 by taking countable unions, countable intersections, and complements. Then it is easy to see that the cardinality of E_1 is $c^{\aleph_0} = c$. Let Ω be the first uncountable ordinal, and $\alpha < \Omega$. Define E_α to be the collection of sets which can be obtained from E_β by taking countable unions, countable intersections, or complements for all ordinal numbers $\beta < \alpha$. Then the cardinality of E_α is again no more than $c^{\aleph_0} \cdot \aleph_0 = c$. Let $E = \bigcup \{E_\alpha : \alpha < \Omega\}$. Then E is a σ-algebra, which is the class of all Borel sets. Since the cardinality of E_α is c for all $\alpha < \Omega$, by transfinite induction, the cardinality of E is $c \cdot \aleph_1 = c$. $\square$

The following proposition says that Borel sets do not exhaust all measurable sets:

3.8. Proposition. *There exists a measurable set which is not Borel.*

Proof. Consider the unit interval $[0, 1]$ and the Cantor ternary set F. Then $m(F) = 0$. Let $A \subset F$ be nonempty. Then A is a set of measure zero, and hence, A is measurable. Therefore, the class of all measurable sets contains the power set of F. Since F has cardinality c, its power set will have cardinality 2^c, which is strictly larger than c. On the other hand, all subsets of $[0, 1]$ are 2^c in number. Therefore, there are exactly 2^c measurable sets, but there are only c Borel sets, which proves that not all measurable sets are Borel. $\square$

The proof that shows there exists a non-Borel measurable set is a cardinal number argument. One can, in fact, construct a non-Borel measurable set. The first real example of such a set is due to the Russian mathematician M.Ya. Suslin (1894–1919), in "Sur une definition des ensembles measurables B sans nombres transfinis" (1917). For a study of Suslin's sets (which are known as *analytic sets*) we refer the interested reader to Kuratowski, *Topology*, I (1966).

EXERCISES 3

A. Show that the following collections of sets are examples of algebras:
   (a) Let $\mathcal{A}$ be the collection of all unions of half-open intervals of the form $[a, b)$.
   (b) Let X be an uncountable set. Let $\mathcal{A}$ be the collection of all sets which either are countable or have countable complements.

B. Is the collection of all open sets in $\mathbb{R}$ an algebra?

C. What is the algebra generated by the collection $\mathcal{C}$ containing only one subset E of X?

D. Let $f: [a, b] \to \mathbb{R}^*$ be measurable. Show that if $\mathcal{C}$ is a collection of subsets E of $\mathbb{R}^*$ for which $f^{-1}(E) = \{x \in [a, b]: f(x) \in E\}$ is measurable, then $\mathcal{C}$ is a σ-algebra.

E. Let $f: [a, b] \to \mathbb{R}^*$ be measurable and let B be a Borel set. Show that $f^{-1}(B)$ is measurable. *Hint*: Exercise D.

§4. Nonmeasurable Sets

Up to now we have proved that certain classes of sets are measurable. This leads to the question whether there exist sets which are not measurable.

The study of this problem has coincided with a period in which the foundations of set theory have been critically examined. There appear to be two obvious ways to solve this problem: Either construct such a set, or else show that the assumption of their nonexistence contradicts some axiom of set theory. It turns out that the proof of the existence of such a set is impossible without using the Axiom of Choice. There had been several examples of nonmeasurable sets given by Vitali (1905), Van Vleck (1908), F. Bernstein (1908), and others, but all of these examples required the use of the axiom of choice in their construction. Lebesgue himself did not admit the nonconstructive methods by which nonmeasurable sets had been produced [see Lebesgue (1926), Appendix in this book]. However, the problem was recently solved by Robert Solovay in his article "A model of set theory in which every set of reals is Lebesgue measurable" (1970). Roughly, his result is that acceptance of the statement "all sets are Lebesgue measurable" as an axiom of set theory is consistent with the usual axioms of set theory if we do not admit the axiom of choice.

We first show the measure m is translation-invariant.

4.1. Proposition. *Let $A \subset [a, b]$ be a measurable set. Then, for every real number r, the set*

$$A + r = \{x + r : x \in A\}$$

is a measurable subset of $[a + r, b + r]$ and $m(A + r) = m(A)$.

This follows from the following proposition:

4.2. Proposition. *For any function $f : [a, b] \to \mathbb{R}$ and any real number r, let $f_r : [a + r, b + r] \to \mathbb{R}$ defined by $f_r(x) = f(x - r)$. If f is integrable on $[a, b]$, then f_r is integrable on $[a + r, b + r]$ and*

$$\int_a^b f(x)\, dx = \int_{a+r}^{b+r} f_r(x)\, dx.$$

Proof of Proposition **4.1.** Let $A \subset [a, b]$ be measurable. Then the characteristic function χ_A is integrable on $[a, b]$ and $m(A) = \int_a^b \chi_A(x)\, dx$. By Proposition 4.2, χ_{A+r} is integrable on $[a + r, b + r]$ and

$$\int_{a+r}^{b+r} \chi_{A+r}(x)\, dx = \int_a^b \chi_A(x)\, dx.$$

Therefore, $m(A + r) = m(A)$.   □

Proof of Proposition **4.2.** We outline the proof. This consists of three easy steps. Show first the proposition is true for step functions, second for functions in the class L^+. Finally, show that the proposition holds for integrable functions. The reader should supply details of this proof (see Exercise 4A).

□

We give below an example of a nonmeasurable set which is a slight modification of one described by the Italian mathematician Guiseppe Vitali (1875–1932), in "Sul problema della misura dei gruppi de punti di una retta" (1905). For a different example we refer the interested reader to Edward B. Van Vleck, "On nonmeasurable sets of points, with an example" (1908).

Before showing such an example, we shall discuss rather informally the Axiom of Choice.

4.3. Axiom of Choice. *Let $\mathscr{C}$ be any collection of nonempty sets. Then there is a function f defined on $\mathscr{C}$ which assigns to each set $A \in \mathscr{C}$ an element $f(A) \in A$.*

The function f is called a *choice function*, and its existence may be thought of as the result of choosing for each of the sets A an element in A. There is no difficulty in doing this if there are only a finite number of sets in $\mathscr{C}$, but we definitely need the axiom in case the collection $\mathscr{C}$ is infinite. It should be noticed that the axiom only asserts the existence of a choice function and is not concerned with the problem of how such a function may be constructed. Because of this fact, some reject it totally. Many accept it without any reservation.

In the following construction of a nonmeasurable set we use a special case of the axiom of choice, namely, the collection $\mathscr{C}$ consisting of mutually disjoint nonempty sets:

4.4. Proposition. *There exists a nonmeasurable set in the interval $[0, 1]$.*

Proof. Let $I = (0, 1)$ and for $x \in I$, let $I(x) = \{\xi \in I: \xi - x \text{ is rational}\}$.

(1) $I(x) = I(y)$ if $x - y$ is rational. For if $\xi \in I(x)$, then $\xi - y = \xi - x + x - y$ is rational since both $\xi - x$ and $x - y$ are rational. Therefore $\xi \in I(y)$. Similarly, if $\xi \in I(y)$, then $\xi \in I(x)$, and hence, $I(x) = I(y)$.

(2) $I(x) \cap I(y) = \varnothing$ if $x - y$ is irrational. For it there is ξ such that $\xi \in I(x) \cap I(y)$, then $x - y = (\xi - y) - (\xi - x)$ is rational, which contradicts the condition that $x - y$ is irrational, so (2) is established.

(3) if $I(x) \cap I(y) \neq \varnothing$, then $I(x) = I(y)$. The follows immediately from (1) and (2).

Let $\mathscr{C}$ denote the collection of all $I(x)$ as x moves over the interval I. Then $\mathscr{C}$ satisfies the hypothesis of the axiom of choice. Let f be a choice function for $\mathscr{C}$ and let A denote the image of f. Then A is a set formed by taking one point from each of the mutually disjoint sets $I(x)$.

Let $r_1, r_2, \ldots$ be all rationals on $(-1, 1)$, and let $A_n = A + r_n = \{x + r_n : x \in A\}$. Then it is easy to see that $A_n \subset (-1, 2)$.

(4) $I \subset \bigcup_{n=1}^{\infty} A_n$. For let $x \in I$; then $x \in I(x)$. Let ξ be the point of $I(x)$ which is in A. Then $|\xi - x| < 1$ and $\xi - x$ is rational. Therefore, there is a rational r_n in $(-1, 1)$ such that $x = \xi + r_n \in A_n$.

(5) $A_m \cap A_n = \varnothing$ if $m \neq n$. For if $x \in A_m \cap A_n$, then $x = \xi + r_m = \eta + r_n$, where $\xi, \eta \in A$. Thus $\xi - \eta = r_n - r_m$ is a rational number, and hence $I(\xi) = I(\eta)$ by (1). Now $(\xi) = A \cap I(\xi) = A \cap I(\eta) = (\eta)$, and hence $\xi = \eta$. But this makes $r_m = r_n$ and $m = n$, which is a contradiction.

We claim that A is not measurable. If A is measurable, then the A_n's are measurable and $m(A_n) = m(A)$ by Proposition 4.1. Since the A_n's are mutually disjoint, then by the countable additivity of the Lebesgue measure m, we have

$$1 = m(I) \le \sum_{n=1}^{\infty} m(A_n) \le m((-1, 2)) = 3,$$

or

$$1 \le \sum_{n=1}^{\infty} m(A) \le 3.$$

This shows that $m(A) \ne 0$ since $1 \le \sum_{n=1}^{\infty} m(A)$, and at the same time $m(A) = 0$ since $\sum_{n=1}^{\infty} m(A) \le 3$. This contradiction leads us to conclude that the set A is nonmeasurable.   □

4.5. Corollary. *There exists a nonmeasurable bounded function defined on* [0, 1].

Proof. Let A be a nonmeasurable set. Then the characteristic function χ_A is a nonmeasurable bounded function.   □

If we had partitioned an arbitrary set of positive measure rather than the open interval (0, 1) into classes $I(x)$, then, repeating the same argument word by word, we would obtain a monmeasurable set A. Therefore, we have the following assertion:

4.6. Proposition. *Every set of positive measure contains a nonmeasurable set.*

EXERCISES 4

A. Prove Proposition 4.2.

B. Let $A \subset [a, b]$ be measurable. Show that the set $-A = \{x: -x \in A\} \subset [-b, -a]$ is measurable and $m(-A) = m(A)$.

C. Let $A \subset [a, b]$ be measurable. For any number $r \ne 0$, let $rA = \{rx: x \in A\}$. Show that rA is measurable and

$$m(rA) = |r| m(A).$$

D. Prove Proposition 4.6.

E. A function $\varphi: \mathbb{R} \to \mathbb{R}$ is called an *isometry* if $|\varphi(x) - \varphi(y)| = |x - y|$ for all x and y.
   (a) Show that if A is a measurable set in $[a, b]$ and if φ is an isometry, then $\varphi(A)$ is measurable in the interval whose endpoints are $\varphi(a)$ and $\varphi(b)$, and

$$m(\varphi(A)) = m(A).$$

   (b) Show that every isometry has one of the following forms:

$$\varphi(x) = x + c \qquad \text{(translation)},$$

$$\varphi(x) = -x \qquad \text{(reflection in the origin)},$$

$$\varphi(x) = -x + c.$$

F. Show that there exists a nonmeasurable set which is dense in $[0, 1]$.

G. Let A be the nonmeasurable set constructed in Proposition 4.4. Show that if E is measurable and $E \subset A$, then $m(E) = 0$. *Hint*: Consider $E_n = E + r_n$.

H. What happens if we replace $+$ and $-$ in the construction of the nonmeasurable set in the proof of Proposition 4.4 by $\cdot$ and $\div$?

I. Let A be a nonmeasurable set in $(0, 1)$. Define $f: (0, 1) \to (0, 2)$ by

$$f(x) = \begin{cases} x + 1 & \text{if } x \in A, \\ x & \text{if } x \notin A. \end{cases}$$

(1) Show that if E is a set of measure zero, then $f(E)$ is of measure zero.
(2) Show that there is a measurable set E in $(0, 1)$ such that $f(E)$ is nonmeasurable.

§5. Structure of Measurable Sets

In this section, among other things, we shall show that every measurable set of positive measure is "almost" a union of finitely many intervals. The word *almost* can be interpreted as *within a set of arbitrarily small measure*. Throughout this section open and closed sets are defined relative to $[a, b]$.

5.1. Proposition. *Let A be a measurable set and let $\varepsilon > 0$ be arbitrary. Then there exists a Borel set B composed of a* finite *number of intervals such that*

$$m((A \setminus B) \cup (B \setminus A)) < \varepsilon.$$

Proof. Since A is measurable, the characteristic function χ_A is measurable, and hence, by Definition 1.1, $\chi_A = \lim \varphi_n$ almost everywhere, where the φ_n's are step functions. We define step functions ψ_n by

$$\psi_n(x) = \begin{cases} 1 & \text{if } \varphi_n(x) \geq \tfrac{1}{2}, \\ 0 & \text{if } \varphi_n(x) < \tfrac{1}{2}. \end{cases}$$

Then the sequence (ψ_n) also converges almost everywhere to χ_A. We observe that ψ_n is a characteristic function of a set which is a finite union of intervals. Let B_n be such that $\chi_{B_n} = \psi_n$. Then it is immediate that

$$(\psi_n - \chi_A)^+ = \chi_{B_n \setminus A} \qquad \text{and} \qquad (\psi_n - \chi_A)^- = \chi_{A \setminus B_n}.$$

Hence

$$|\psi_n - \chi_A| = \chi_{B_n \setminus A} + \chi_{A \setminus B_n}$$

and

$$\lim \int_a^b |\psi_n(x) - \chi_A(x)|\, dx = 0.$$

(Why?) Therefore, for $\varepsilon > 0$ there exists a natural number n_0 such that

$$\int_a^b |\psi_{n_0}(x) - \chi_A(x)|\, dx < \frac{\varepsilon}{2}.$$

For this particular n_0, we have

$$m(A \setminus B_{n_0}) + m(B_{n_0} \setminus A) = \int_a^b [\chi_{A \setminus B_{n_0}}(x) + \chi_{B_{n_0} \setminus A}(x)] \, dx$$

$$= \int_a^b |\psi_{n_0}(x) - \chi_A(x)| \, dx < \frac{\varepsilon}{2},$$

which shows that $m(A \setminus B_{n_0}) < \varepsilon/2$ and $m(B_{n_0} \setminus A) < \varepsilon/2$. Let $B = B_{n_0}$. Then $m((A \setminus B) \cup (B \setminus A)) < \varepsilon$.   □

5.2. Proposition. *Let A be a measurable set. Then for each $\varepsilon > 0$ there exists an open set G containing A such that*

$$0 \le m(G) - m(A) < \varepsilon.$$

Proof. For each real number $\varepsilon > 0$ and any natural number n, by Proposition 5.1, we can find an open set B_n such that

$$m((A \setminus B_n) \cup (B_n \setminus A)) < \frac{\varepsilon}{2^{n+1}}.$$

Let $B = \bigcup_{n=1}^\infty B_n$, $C = \bigcap_{n=1}^\infty (A \setminus B_n)$, and $D = \bigcup_{n=1}^\infty (B_n \setminus A)$. Then

$$m(C) = \lim m(A \setminus B_n) = 0$$

and

$$m(D) \le \sum_{n=1}^\infty m(B_n \setminus A) < \sum_{n=1}^\infty \frac{\varepsilon}{2^{n+1}} = \frac{\varepsilon}{2}.$$

We claim that $A \subset B \cup C$. For, if $x \in A$, then either $x \in B_n$ or $x \in A \setminus B_n$ for each n since $A \subset B_n \cup (A \setminus B_n)$. If $x \notin B$, then $x \notin B_n$ for all n, and hence $x \in A \setminus B_n$ for all n, which implies $x \in C$. Similarly, if $x \notin C$, then $x \in B$. This proves $A \subset B \cup C$. We also have $B \subset A \cup D$ (why?). Since $m(C) = 0$, we can find an open set C' containing C with $m(C') < \varepsilon/2$. Let $G = B \cup C'$. Then G is an open set containing A. Then

$$m(G) \le m(B) + m(C') \le m(A) + m(D) + m(C') < m(A) + \varepsilon.$$

Therefore, $0 \le m(G) - m(A) < \varepsilon$.   □

5.3. Proposition. *Let A be a measurable set. Then there exists a decreasing sequence (G_n) of open sets containing A such that*

$$m(A) = \lim m(G_n).$$

Proof. For each n, we can find an open set U_n containing A such that

$$0 \le m(U_n) - m(A) < \frac{1}{n}.$$

Let $G_n = U_1 \cap \cdots \cap U_n$. Then G_n is open and $m(A) = \lim m(G_n)$.   □

Apply the previous results to $[a, b] \setminus A$ to obtain the following proposition:

5.4. Proposition. *Let A be a measurable set. Then for each $\varepsilon > 0$ there exists a closed set F contained in A such that*

$$0 \leq m(A) - m(F) < \varepsilon.$$

5.5. Proposition. *Let A be a measurable set. Then there exists an increasing sequence (F_n) of closed sets contained in A such that*

$$m(A) = \lim m(F_n).$$

5.6. Proposition. *If A is measurable, then there exist Borel sets B_1 and B_2 such that $B_1 \supset A \supset B_2$ and $m(B_1) = m(A) = m(B_2)$.*

Now we will prove the following criterion for measurability of sets:

5.7. Proposition. *If A is a measurable set, then*

$$m(A) = \inf\{m(G): G \supset A; G \text{ open}\} = \sup\{m(F): F \subset A; F \text{ closed}\}.$$

Conversely, if $\inf\{m(G): G \supset A; G \text{ open}\} = \sup\{m(F): F \subset A; F \text{ closed}\}$, then A is measurable.

Proof. The first part is an easy consequence of Propositions 5.3 and 5.5. We show the second part. Let (G_n) be a monotone decreasing sequence of open sets containing A such that $\lim m(G_n) = \inf\{m(G): G \supset A; G \text{ open}\}$, and let (F_n) be a monotone increasing sequence of closed sets contained in A such that $\lim m(F_n) = \sup\{m(F): F \subset A; F \text{ closed}\}$. Then $\lim m(G_n) = m(\bigcap_{n=1}^{\infty} G_n)$ and $\lim m(F_n) = m(\bigcup_{n=1}^{\infty} F_n)$. But $\bigcup_{n=1}^{\infty} F_n \subset A \subset \bigcap_{n=1}^{\infty} G_n$, and so

$$m\left(\bigcap_{n=1}^{\infty} G_n \setminus A\right) \leq m\left(\bigcap_{n=1}^{\infty} G_n \setminus \bigcup_{n=1}^{\infty} F_n\right) = 0.$$

Thus A is measurable since it differs from $\bigcap_{n=1}^{\infty} G_n$ by a set of measure zero. $\square$

EXERCISES 5

A. Prove Proposition 5.4.

B. Prove Proposition 5.5.

C. Show that a set A is measurable if and only if, for every $\varepsilon > 0$, there exist both an open set G and a closed set F such that $F \subset A \subset G$ and

$$m(G \setminus F) < \varepsilon$$

D. Let $A \subset [a, b]$ be such that

$$m(A) > \tfrac{1}{2}(b - a).$$

Show that A contains a subset of positive measure which is symmetric to the midpoint of $[a, b]$. *Hint*: Consider the intersection of A with its reflection through the midpoint.

§6. More About Measurable Functions

In this section we will show that measurable sets and measurable functions are closely related. Our purpose is to characterize measurable functions in terms of measurable sets.

6.1. Proposition. *If f is a measurable function and c is a constant, then the set*

$$\{x \in [a, b]: f(x) \le c\}$$

is measurable.

Proof. We introduce the function f_c, which is the function f truncated below by c:

$$f_c(x) = \max\{f(x), c\}.$$

For $h > 0$, consider the quotient

$$\frac{f_{c+h}(x) - f_c(x)}{h} = \begin{cases} 0 & \text{if } f(x) \ge c + h, \\ 1 & \text{if } f(x) \le c. \end{cases}$$

This quotient is measurable. Its limit as $h \to 0$ is the characteristic function of the set $\{x \in [a, b]: f(x) \le c\}$, which is also measurable, by Proposition 1.6. (More precisely, replace h by $1/n$, $n \in \mathbb{N}$, and let $n \to \infty$.) □

6.2. Proposition. *The following statements are equivalent:*

(a) $\{x \in [a, b]: f(x) \le c\}$ *is measurable;*
(b) $\{x \in [a, b]: f(x) > c\}$ *is measurable;*
(c) $\{x \in [a, b]: f(x) \ge c\}$ *is measurable; and*
(d) $\{x \in [a, b]: f(x) < c\}$ *is measurable.*

Proof.

$$\{x \in [a, b]: f(x) > c\} = [a, b] \setminus \{x \in [a, b]: f(x) \le c\},$$

$$\{x \in [a, b]: f(x) \ge c\} = \bigcap_{n=1}^{\infty} \left\{x \in [a, b]: f(x) > c - \frac{1}{n}\right\},$$

$$\{x \in [a, b]: f(x) < c\} = [a, b] \setminus \{x \in [a, b]: f(x) \ge c\},$$

$$\{x \in [a, b]: f(x) \le c\} = \bigcap_{n=1}^{\infty} \left\{x \in [a, b]: f(x) < c + \frac{1}{n}\right\}.$$ □

Combining the above two propositions we can characterize a measurable function in the following way:

6.3. Proposition. *Let f be a function defined on $[a, b]$. Then f is measurable if and only if $\{x \in [a, b]: f(x) \le c\}$ is measurable for each constant c.*

Proof. Suppose that $\{x \in [a, b]: f(x) \le c\}$ is measurable for each constant c. But then, since

$$\{x \in [a, b]: c \le f(x) < d\}$$

$$= \{x \in [a, b]: f(x) < d\} \setminus \{x \in [a, b]: f(x) < c\},$$

the set $\{x \in [a, b]: c \le f(x) < d\}$ is measurable. Next, let n be a fixed integer and let k be any integer. Define a function f_n by

$$f_n(x) = \frac{k}{2^n} \quad \text{if} \quad \frac{k}{2^n} \le f(x) < \frac{k+1}{2^n}.$$

The function f_n is the sum of a convergent series of measurable functions and hence it is measurable by Proposition 1.6. In fact,

$$f_n(x) = \sum_{k=-\infty}^{\infty} \frac{k}{2^n} \chi_{E_k^n}(x),$$

where

$$E_k^n = \left\{x \in [a, b]: \frac{k}{2^n} \le f(x) < \frac{k+1}{2^n}\right\}.$$

It is clear that the inequality

$$|f_n(x) - f(x)| < \frac{1}{2^n}$$

holds. Therefore, $f = \lim f_n$, which is also measurable by Proposition 1.6. $\square$

The above characterization of a measurable function is extremely useful and is frequently given as the definition. The reader should recognize that the notion of a measurable function characterized in Proposition 6.3 is quite analogous to that of a continuous function.

EXERCISES 6

A. Show that a function f on $[a, b]$ is measurable if and only if the set $\{x \in [a, b]: f(x) > r\}$ is measurable for each rational number r.

B. If f and g are measurable functions, show that $\{x \in [a, b]: f(x) \le g(x)]$ is a measurable set.

C. Show that if f is a measurable function and if $g = f$ almost everywhere, then g is a measurable function.

D. If f is such that $|f|$ is measurable, does f have to be measurable?

E. Show that the function h defined in Example 5.5 is integrable.

§7. Egoroff's Theorem

Mathematical analysis is heavily concerned with approximation of complicated functions by means of simple functions. In a sense, it is a main goal of analysis. Up to this point we have seen that any measurable function is approximated by step functions. The convergence occurring in this approximation has been almost everywhere pointwise convergence. In this section we introduce some other kinds of convergence and prove theorems which indicate their relaionships.

We recall the concept of uniform convergence of a sequence of functions. A sequence (f_n) of functions on $A \subset \mathbb{R}$ *converges uniformly* on A to a function f if for each $\varepsilon > 0$ there is a natural number N (depending on ε but not on $x \in A$) such that, for all $n \geq N$ and $x \in A$, we have

$$|f_n(x) - f(x)| < \varepsilon.$$

The following example illustrates the difference between pointwise and uniform convergence. Let (f_n) be a sequence of functions defined on $[0, 1]$ by

$$f_n(x) = x^n.$$

The (f_n) converges to the function f, $f(x) = 0$ if $0 \leq x < 1$ and $f(x) = 1$ if $x = 1$, pointwise. But it is easy to show that the convergence is not uniform. However, (f_n) converges to 0 uniformly on the closed interval $[0, 1 - \varepsilon]$ for each $\varepsilon > 0$. This motivates the following concept:

7.1. Definition. A sequence (f_n) of functions defined on $[a, b]$ *converges almost uniformly* to a function f on a measurable set A if for each $\varepsilon > 0$ there is a measurable set $E \subset A$ such that $m(E) < \varepsilon$ and the sequence (f_n) converges to f uniformly on $A \backslash E$.

It is clear that uniform convergence implies almost uniform convergence, but not conversely. The above example shows this fact.

The following theorem is of fundamental importance in the study of convergence of measurable functions. It is due to the Russian mathematician D.F. Egoroff (1869–1931), "Sur les suites de fonctions mesurables" (1911). This theorem establishes an interesting relationship between convergence almost everywhere and uniform convergence.

7.2. Theorem (Egoroff, 1911). *Let (f_n) be a sequence of measurable functions on $[a, b]$ which converges almost everywhere to a function f. Then (f_n) converges almost uniformly to f on $[a, b]$.*

Proof. Since the limit f is also measurable by Proposition 1.6, the functions $f - f_n$ are measurable. Therefore this problem can be reduced to the case of a sequence that converges to zero by replacing f_n by $f - f_n$ if necessary. Therefore, we may assume from the beginning that the functions f_n are non-negative and converge to zero monotonically. If this is not the case, we can replace f_n by

$$\sup\{|f_n|, |f_{n+1}|, \ldots\}.$$

Therefore, assume that $f_n \downarrow 0$ almost everywhere.

Let N be the subset of $[a, b]$ of measure zero on which (f_n) does not converge to zero. For each pair of natural numbers k, n let

$$E_{kn} = \{x \in [a, b] \backslash N : 0 \leq f_m(x) < 1/2^k \text{ if } m \geq n\}.$$

Then these sets are measurable and $E_{kn} \subset E_{k,n+1}$. Since (f_n) converges to zero on $[a, b] \backslash N$, we see that for each fixed k we have

$$\bigcup_{n=1}^{\infty} E_{kn} = [a, b] \backslash N.$$

Then, by Proposition 2.5, we have

$$\lim m(E_{kn}) = m([a, b] \backslash N) = b - a.$$

Consequently, given $\varepsilon > 0$, there is a natural number $n(k)$ such that

$$|m(E_{k, n(k)}) - (b - a)| < \frac{\varepsilon}{2^k}$$

or

$$m([a, b] \backslash E_{k, n(k)}) < \frac{\varepsilon}{2^k}.$$

Let

$$E = [a, b] \backslash \bigcap_{k=1}^{\infty} E_{k, n(k)} = \bigcup_{k=1}^{\infty} [a, b] \backslash E_{k, n(k)}.$$

We see that

$$m(E) \leq \sum_{k=1}^{\infty} \frac{\varepsilon}{2^k} = \varepsilon.$$

Now, $[a, b] \backslash E = \bigcap_{k=1}^{\infty} E_{k, n(k)}$, and so, if $x \in [a, b] \backslash E$, we have

$$0 \leq f_m(x) < \frac{1}{2^k} \quad \text{if } m \geq n(k),$$

this being valid for $k = 1, 2, \ldots$. Therefore, (f_n) converges to zero uniformly on $[a, b] \backslash E$. $\square$

It is not difficult to see that in the statement of the theorem the closed interval $[a, b]$ can be replaced by any measurable set (see Exercise 7B).

7.3 Corollary. *Every measurable function f on $[a, b]$ is an almost uniform limit of a sequence of step functions defined on $[a, b]$.*

The converse of Theorem 7.2 is also true.

7.4. Proposition. *If (f_n) is a sequence of measurable functions on $[a, b]$ which converges to f almost uniformly, then (f_n) converges to f almost everywhere.*

Proof. For each n, let $E_n \subset [a, b]$ be a measurable set such that $m(E_n) < 1/n$ and the sequence (f_n) converges to f uniformly on $[a, b] \setminus E_n$. If $E = \bigcap_{n=1}^{\infty} E_n$, then

$$m(E) \leq m(E_n) < \frac{1}{n},$$

so that $m(E) = 0$, and it is easy to see that (f_n) converges to f on $[0, 1] \setminus E$. $\qquad\square$

The following is an easy consequence of the Egoroff theorem:

7.5. Proposition. *Let (f_n) be a sequence of measurable functions on $[a, b]$ which converges to f almost everywhere. Then for each $\varepsilon > 0$,*

$$\lim m(\{x \in [a, b]: |f_n(x) - f(x)| \geq \varepsilon\}) = 0.$$

Proof. Corresponding to an arbitrarily assigned $\delta > 0$, there exists a measurable set $E \subset [a, b]$ such that $m(E) < \delta$ and such that (f_n) converges to f uniformly on $[a, b] \setminus E$. Then, given $\varepsilon > 0$, choose N so that $|f_n(x) - f(x)| < \varepsilon$ if $n \geq N$ and $x \in [a, b] \setminus E$. We see that

$$\{x \in [a, b]: |f_n(x) - f(x)| \geq \varepsilon\} \subset E \qquad \text{if} \quad n \geq N.$$

Therefore,

$$m(\{x \in [a, b]: |f_n(x) - f(x)| \geq \varepsilon\}) < \delta \qquad \text{if} \quad n \geq N.$$

This proves that

$$\lim m(\{x \in [a, b]: |f_n(x) - f(x)| \geq \varepsilon\}) = 0. \qquad\square$$

This proposition motivates the following definition:

7.6. Definition. A sequence (f_n) of measurable functions is said to *converge in measure* to a measurable function f if, for every $\varepsilon > 0$,

$$\lim m(\{x \in [a, b]: |f_n(x) - f(x)| \geq \varepsilon\}) = 0.$$

In accordance with this notion, Proposition 7.5 states that *if (f_n) converges to a function f almost everywhere, then (f_n) converges in measure to f.* This concept was introduced and studied by F. Riesz and E. Fischer in 1906–1907. One of its most important uses is in connection with the completeness of the Lebesgue space (see Exercise 7F).

Much of the usefulness of the concept of convergence in measure lies in the fact that the space $\mathcal{M}$ of measurable functions on $[a, b]$ which are finite almost everywhere is a Banach space, as described in Exercise 7H. On $\mathcal{M}$, the notion of convergence of functions corresponds to convergence in measure.

The following example shows that the full converse of Proposition 7.5 is not true:

7.7. Example. Let

$$E_{nk} = \left[\frac{k-1}{n}, \frac{k}{n}\right], \qquad k = 1, 2, \ldots, n,$$

and let (E_n) be the sequence

$$E_{11}, \quad E_{21}, \quad E_{22}, \quad E_{31}, \quad E_{32}, \quad E_{33}, \ldots.$$

Let $f_n = \chi_{E_n}$. Then (f_n) converges in measure to 0, since $m(E_n) \to 0$ as $n \to \infty$ and $\{x \in [a, b]: |f_n(x)| > 0\} = E_n$. But for each x, $f_n(x) = 1$ for infinitely many values of n, and hence $f_n(x) \to 0$ is always false.

Although the full converse of Proposition 7.5 is not true, we have the following:

7.8. Proposition (Riesz, 1909). *If (f_n) is a sequence of measurable functions on $[a, b]$ which converges in measure to f, then there exists a subsequence (f_{n_k}) of (f_n) which converges to f almost everywhere.*

Proof. Let (ε_n) be a sequence of positive real numbers which converges to 0, and let (δ_n) be a sequence of positive numbers such that $\sum_{n=1}^{\infty} \delta_n < \infty$. Since (f_n) converges in measure to f, we can find an index n_1 such that

$$m(\{x \in [a, b]: |f_{n_1}(x) - f(x)| \geq \varepsilon_1\}) < \delta_1.$$

In general, by induction, we can choose n_k such that

$$n_1 < n_2 < \cdots < n_{k-1} < n_k,$$

and

$$m(\{x \in [a, b]: |f_{n_k}(x) - f(x)| \geq \varepsilon_k\}) < \delta_k.$$

We shall show that the subsequence (f_{n_k}) converges to f almost everywhere. In fact, let

$$E_k = \bigcup_{i=k}^{\infty} \{x \in [a, b]: |f_{n_i}(x) - f(x)| \geq \varepsilon_i\},$$

and

$$E = \bigcap_{k=1}^{\infty} E_k.$$

Then $m(E) = \lim m(E_k) = \lim \sum_{i=k}^{\infty} \delta_i = 0$. It remains to verify that $f_{n_k} \to f$

for all $x \in [a, b] \backslash E$. Let $x_0 \in [a, b] \backslash E$. Then there is an N such that $x_0 \notin E_N$. Then

$$x_0 \notin \{x \in [a, b] : |f_{n_k}(x) - f(x)| \geq \varepsilon_k\}$$

for all $k \geq N$, i.e.,

$$|f_{n_k}(x) - f(x)| < \varepsilon_k.$$

Since $\varepsilon_k \to 0$, it is clear that

$$f_{n_k}(x_0) \to f(x_0).$$

This completes the proof.   □

Another application of the Egoroff theorem is the following theorem known as Lusin's theorem, after the Russian mathematician N.N. Lusin (1883–1952), "Sur les propriétes des fonctions measurables" (1912). The theorem claims that a measurable function looks "almost" like a continuous function.

7.9. Theorem (Lusin, 1912). *A function f defined on the closed interval $[a, b]$ is measurable if and only if for every $\varepsilon > 0$ there exists a measurable set E with $m(E) < \varepsilon$ such that f is continuous on $[a, b] \backslash E$.*

Proof. Let (φ_n) be a sequence of step functions converging almost everywhere to f. Let N be the set of points of divergence of the sequence (φ_n) and the points of discontinuities of φ_n. Then N is a set of measure zero. We can cover N by a countable system of open intervals I_n of overall length $< \varepsilon/2$. By Egoroff's theorem, we can find a measurable set E with $m(E) < \varepsilon/2$ such that (φ_n) converges to f uniformly on $[a, b] \backslash E$. We then have a set $F = [a, b] \backslash (E \cup \bigcup_{n=1}^{\infty} I_n)$ and (φ_n) converges to f uniformly on F; hence f is continuous as the uniform limit of a sequence of continuous functions on F.

We leave a proof of the converse statement to the reader as Exercise 7G.   □

Before closing this section we summarize the relations between the different kinds of convergence of sequences of measurable functions on $[a, b]$. This can be done by the following diagram: the arrows indicate implication in the diagram:

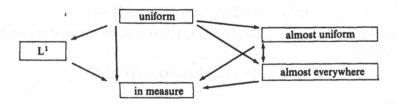

where L^1 convergence means convergence with respect to the L^1 norm. Exercise 7E shows that L^1 convergence implies convergence in measure.

EXERCISES 7

A. Show that the sequence (f_n) of functions on $[0, 1]$ defined by

$$f_n(x) = x^n$$

converges almost uniformly without using the Egoroff theorem.

B. In the statement of the Egoroff theorem show that the closed interval $[a, b]$ can be replaced by any bounded measurable set.

C. The Egoroff theorem does not claim that there exists a subset $E \subset [a, b]$ with $m(E) = 0$ and that the sequence (f_n) converges to f uniformly on $[a, b] \backslash E$. However, prove that there exists a sequence (E_n) of measurable sets in $[a, b]$ such that $m([a, b] \backslash \bigcup_{n=1}^{\infty} E_n) = 0$ and (f_n) converges to f uniformly on each E_n.

D. Show that if (f_n) converges in measure to f, then (f_n) converges in measure to g whenever $f = g$ almost everywhere.

E. Let (f_n) be a sequence of integrable functions which converges to a function f with respect to the L^1 norm. Show that (f_n) converges to f in measure.

F. Use Proposition 7.8 to prove that $L^1[a, b]$ is complete.

G. Prove that if for each $\varepsilon > 0$ there exists a measurable set E with $m(E) < \varepsilon$ such that f is continuous on $[a, b] \backslash E$, then f is measurable on $[a, b]$.

H. M. Frechet (1928). Let $\mathcal{M}$ be the class of measurable functions on $[a, b]$ which are finite almost everywhere. If f is in $\mathcal{M}$, the function

$$\frac{|f|}{1 + |f|}$$

is well defined almost everywhere and belongs to $\mathcal{M}$. Since it is bounded, we may consider the integral

$$\|f\| = \int_a^b \frac{|f|}{1 + |f|} \, dx.$$

(a) Show that $\|f\| = 0$ if and only if $f \sim 0$; i.e., $f = 0$ almost everywhere.
(b) Show that $\|f_n - f\| \to 0$ as $n \to \infty$ if and only if $f_n \to f$ in measure. Let $\bar{\mathcal{M}}$ be the set of equivalence classes in $\mathcal{M}$ determined by the equivalence relation $\sim$.
(c) Make $\bar{\mathcal{M}}$ into a complete metric space.

I. Let (f_n) be a sequence of measurable functions on $[a, b]$. Show that the set E of points where $\lim f_n(x)$ exists is measurable.
   Hint: $E = \bigcap_{k=1}^{\infty} \bigcup_{n=1}^{\infty} \bigcap_{m=1}^{\infty} \{x : |f_n(x) - f_m(x)| < 1/k\}$.

J. For each natural number n, let $f_n : [0, 2] \to \mathbb{R}$ be defined by

$$f_n(x) = \begin{cases} \sqrt{n} & \text{if } 1/n \le x \le 2/n, \\ 0 & \text{otherwise.} \end{cases}$$

Show that

$$\int_0^2 |f_n(x)| \, dx \to 0$$

but (f_n) does not converge to 0 uniformly on $[0, 2]$.

K. Let $f_n: \mathbb{R} \to \mathbb{R}$ be defined by

$$f_n(x) = \begin{cases} 1/n & \text{if } |x| \leq n, \\ 0 & \text{if } |x| > n. \end{cases}$$

Then $\int_{\mathbb{R}} |f_n(x)| \, dx = 2$; however, $f_n(x) \to 0$ as $n \to \infty$.

§8. Steinhaus' Theorem

In his paper, "Sur les distances des points dans les ensembles de mesure positive" (1920), the Polish mathematician Hugo Steinhaus showed that *if A is a set of positive measure, then the set*

$$A - A = \{x - y: x \in A \text{ and } y \in A\}$$

contains an open interval. In this section we will present an elementary proof of this theorem and also generalize its result to the sets

$$A + B = \{x + y: x \in A \text{ and } y \in B\},$$

and

$$A - B = \{x - y: x \in A \text{ and } y \in B\}.$$

8.1. Lemma. *Suppose A is a bounded set of real numbers whose Lebesgue measure is positive. Then for each real number k such that $0 \leq k < 1$ there exists an open interval I such that $m(A \cap I) > km(I)$.*

Proof. Suppose the lemma is false. Then for some k such that $0 \leq k < 1$ and for every open interval I we have $m(A \cap I) \leq km(I)$. Let G be a bounded open set containing A. G can be written as a union of countably many disjoint open intervals I_m; i.e., $G = \bigcup_{m=1}^{\infty} I_m$ (see Proposition 5.3, Chapter Zero). Since the measure m is countably additive, $m(G) = \sum_{m=1}^{\infty} m(I_m)$. Then

$$m(A) = m(A \cap G) = \sum_{m=1}^{\infty} m(A \cap I_m) \leq k \sum_{m=1}^{\infty} m(I_m) = km(G).$$

But then, since

$$m(A) = \inf\{m(G): G \text{ is open}, A \subset G\},$$

we have

$$m(A) \leq km(A)$$

and hence $1 \leq k$. This contradicts the fact that $0 \leq k < 1$, and the lemma is proved. $\square$

8.2. Theorem (Steinhaus, 1920). *If A is a set of positive measure, then $A - A$ contains an open interval.*

Proof. Suppose that the set $A - A$ does not contain an open interval. Since A is bounded, we can find an open interval (a, b) such that $m(A \cap (a, b)) > 3(b - a)/4$ by the lemma. Let $B = A \cap (a, b)$. Choose a point p in the complement of $A - A$ satisfying $0 < p < d$, where $d = (b - a)/2$. Consider $C = \{p\} + B$. Then C is a subset of $(a, b + d)$. Therefore, both B and C are in $(a, b + d)$, and $B \cap \quad = \emptyset$. In fact, if $x \in B \cap C$, then $x = b = p + b'$ and hence $p = b - b' \in B - B \subset A - A$, which contradicts the fact that $p \notin A - A$. Now

$$m(B \cup C) = m(B) + m(C) = 2m(B) = 2m(A \cap (a, b)) > \tfrac{3}{2}(b - a) = 3d.$$

But $m(B \cup C) \leq m(a, b + d) = 3d$. Therefore, $3d < 3d$, an obvious contradiction, and the theorem is proved. □

A parallel proof gives the following result:

8.3. Theorem. *If A is a set of positive measure, then $A + A$ contains an open interval.*

Proof. Suppose $A + A$ does not contain an open interval. Let B be such that $B = A \cap (a, b)$ and $m(B) > 3(b - a)/4$ as in the proof of Theorem 8.2. Choose a point p in the complement of $A + A$ such that $a + b < p < a + b + d$, where $d = (b - a)/2$. Consider $C = \{p\} - B$. Then C is a subset of $(a, b + d)$ and $B \cap C = \emptyset$. Now

$$m(B \cup C) = m(B) + m(C) = 2m(B) > \tfrac{3}{2}(b - a) = 3d.$$

But $m(B \cup C) \leq 3d$. This is absurd, and hence the supposition that $A + A$ does not contain an open interval is false. □

To generalize the Steinhaus theorem we need the following lemma:

8.4. Lemma. *If I is an open interval such that $m(A \cap I) > km(I)$ and if n is a natural number, then we can find an open interval $J \subset I$ such that $m(I) = nm(J)$ and $m(A \cap J) > km(J)$.*

Proof. Partition the open interval I into n equal intervals; then at least one of these intervals, say J, should satisfy the relation $m(A \cap J) > km(J)$. □

Now we state a generalized Steinhaus theorem.

8.5. Theorem. *If A and B have positive Lebesgue measure, then $A + B$ contains an open interval.*

Proof. Let $k = \tfrac{3}{4}$. By Lemma 8.1 we can find two open intervals (a, b) and (c, d) such that

$$m(A \cap (a, b)) > k(b - a) \qquad \text{and} \qquad m(B \cap (c, d)) > k(d - c).$$

Then choose a pair of natural numbers m and n satisfying

$$k < \frac{m(d - c)}{n(b - a)} < 1.$$

By the second part of Lemma 8.4, we can find open intervals (a', b') and (c', d') such that

$$m(b' - a') = b - a \quad \text{and} \quad n(d' - c') = d - c,$$

and

$$m(A \cap (a', b')) > k(b' - a'),$$
$$m(B \cap (c', d')) > k(d' - c').$$

Then

$$k < \frac{d' - c'}{b' - a'} < 1.$$

Now let $A' = A \cap (a', b')$ and $B' = B \cap (c', d')$. Then $A' + B' \subset A + B$.

Suppose that $A + B$ does not contain an interval of positive length. Then we can pick a point p in the complement of $A + B$ satisfying

$$\frac{a' + b' + c' + d'}{2} < p < b' + c'.$$

Consider $C = \{p\} - B'$. Then $m(C) = m(B')$. Furthermore, $C \cap A' = \varnothing$, and both A' and C are subsets of (a', b'). Therefore, $m(A' \cup C) \leq b' - a'$. On the other hand,

$$m(A' \cup C) = m(A') + m(C) = m(A') + m(B')$$
$$> k[(b' - a') + (d' - c')]$$
$$> k[(b' - a') + k(b' - a')] > b' - a'.$$

This is absurd, and hence the supposition that $A + B$ does not contain an open interval is false. $\square$

8.6. Corollary. *If A and B have positive Lebesgue measure, then $A - B$ contains an interval of positive length.*

Proof. Let $-B = \{x: -x \in B\}$. Then $m(-B) = m(B) > 0$. But $A - B = A + (-B)$. Therefore, $A - B$ contains an open interval. $\square$

We should notice that the Cantor ternary set F satisfies the result of the Steinhaus theorem although it is a set of measure zero. However, the condition $m(A) > 0$ in the theorem cannot be weakened. For example, if A is a countable set, then $m(A) = 0$ and $A - A$ cannot contain any interval.

EXERCISES 8

A. Show that for the Cantor ternary set F, $F - F = [-1, 1]$. *Hint:* Let $d \in [-1, 1]$. Then the line $y = x + d$ meets one of the squares in $([0, 1/3] \cup [2/3, 1]) \times ([0, 1/3] \cup [2/3, 1])$. Draw these squares and visualize $F \times F$.

B. For a given $A \subset [0, 1]$, let $\varepsilon > 0$ be such that $\varepsilon \leq m(A)$. For $n > 2/\varepsilon$, let $p_1, \ldots, p_n$ be arbitrary numbers in $[0, 1]$. Show that A contains a pair of points whose distance is equal to the distance between some pair of $p_1, \ldots, p_n$. *Hint:* $p_1 + A, \ldots, p_n + A$ cannot be mutually disjoint.

C. Using the preceding problem, show that every set of positive measure contains infinitely many pairs of points at rational distances.

D. **Theorem.** *Let A and B be two measurable subsets of reals with positive measure and let G be an open set in $\mathbb{R}^2$ containing $A \times B$. If $f: G \to \mathbb{R}$ is continuously differentiable and $\partial f/\partial x$, $\partial f/\partial y$ are never zero on G, then $f(A \times B)$ contains an interval.* (See Chae and Peck, 1973. Proof requires the Implicit Function Theorem.)

E. Let (a, b) be as in the proof of the Steinhaus Theorem 8.2. Show that the open interval $(-c, c)$ is entirely contained in $A - A$, where $c = (b - a)/2$.

§9. The Cauchy Functional Equation

In this section, as an application of the Steinhaus theorem, we discuss the Cauchy functional equation

$$f(x + y) = f(x) + f(y)$$

defined for all x, y in $\mathbb{R}$. Such a function is called *additive*. In 1821, in his *Cours d'Analyse*, Cauchy showed that the only continuous additive functions f are those which are linear; i.e., $f(x) = mx$, where m is a constant. In many undergraduate courses such as advanced calculus or intermediate analysis, this fact is proved by showing that such a function should be linear for $x \in \mathbb{Q}$; i.e.,

$$f(x) = mx$$

for all rational x. (In what follows we shall use this assertion.) Then the difference $f(x) - mx$ is continuous and vanishes on rational x, and hence everywhere. In the following development we shall show that if f is measurable in some interval and satisfies the Cauchy functional equation, then f is linear. This result was first proved by the French mathematician Maurice Fréchet in "Pri la funkcia ekvacio" (1913). We shall also demonstrate how complicated the graph of a nonmeasurable function can be.

All functions in this section are defined on the entire real line $\mathbb{R}$.

9.1. Proposition. *Suppose that f is additive and bounded on an interval. Then f is linear.*

Proof. Suppose that f is bounded on $[a, b]$. Then for all y in $[a, b]$, $|f(y)| < M$. If x is in $[0, b - a]$, then $x + a$ is in $[a, b]$, so that from

$$f(x) = f(x + a) - f(a),$$

we get

$$|f(x)| < 2M.$$

Accordingly, if $b - a = c$, f is bounded on $[0, c]$ by $2M$. Also notice that $|f(x)| < 2M$ for $x \in [-c, c]$.

Let x be any real number. Then for any natural number n, we can find a rational r such that $|x - r| < c/n$. Then

$$|f(x) - xf(1)| = |f(x - r) + (r - x)f(1)| \leq \frac{2M + c|f(1)|}{n}.$$

Since this is true for any n, we conclude that

$$f(x) = xf(1) \qquad \text{for all} \quad x \in \mathbb{R}. \qquad \square$$

We now replace an interval in the preceding proposition with a set of positive measure.

9.2. Proposition. *Suppose that f is additive and bounded on a set A of positive measure. Then f is linear.*

Proof. Let I be an open interval such that $I \subset A - A$ and f is bounded on A; say $|f(x)| < M$ for all x in A. Then for any x in I we have $x = a - b$ for some a, b in A. Then

$$|f(x)| = |f(a) - f(b)| \leq |f(a)| + |f(b)| < 2M,$$

so that f is bounded on I. Therefore f is linear by Proposition 9.1. $\qquad \square$

Notice that the assumption $m(A) > 0$ in Proposition 9.2 was made only in order to justify that $A - A$ contains an interval. Hence, we have the following result:

9.3. Corollary. *Suppose that f is additive and bounded on a set A for which $A - A$ contains an interval. Then f is linear. In particular, every additive function which is bounded on the Cantor ternary set is linear.*

We now show that a discontinuous additive function cannot be measurable.

9.4. Proposition. *Suppose that f is additive and measurable in some interval. Then f is linear.*

Proof. Let f be measurable on $[a, b]$. Then the set

$$\{x \in [a, b]: |f(x)| < n\}$$

has positive measure if n is large enough. Therefore f is linear by Proposition 9.2. $\qquad\square$

9.5. Corollary. *Every discontinuous solution of the Cauchy functional equation is not measurable in any interval.*

Based on Corollary 9.5, we shall point out the highly pathological behavior of a nonmeasurable function, namely, a discontinuous additive function.

9.6. Proposition. *Suppose that f is a discontinuous additive function. Then the graph $\{(x, f(x)): x \in \mathbb{R}\}$ is dense in the plane; i.e., every circle contains a point $(x, f(x))$.*

Proof. Let $m = f(1)$. Since f is not linear, there is a point p such that

$$f(p) \neq mp.$$

The graph of f contains all points of the form $(r + sp, rm + sf(p))$, where r and s are rational, since

$$f(r + sp) = rf(1) + sf(p) = rm + sf(p).$$

But

$$(r + sp, rm + sf(p)) = (r, s)\begin{pmatrix} 1 & m \\ p & f(p) \end{pmatrix}.$$

Let M be the matrix

$$\begin{pmatrix} 1 & m \\ p & f(p) \end{pmatrix}.$$

Then M is not singular since $\det M \neq 0$. Hence M is a one–one and continuous mapping of the plane $\mathbb{R}^2$ onto itself. Therefore, M maps dense sets into dense sets. In particular, the image of the set $\{(r, s): r, s \text{ are rational}\}$ under M is dense in the plane. But this image is a subset of the graph of f. Thus the graph of f is dense in the plane $\mathbb{R}^2$. $\qquad\square$

The existence of a discontinuous additive function is dependent on the axiom of choice since such a function is nonmeasurable (Solovay, 1970). For the material in this section, reference might be made to the article Wilansky (1967).

EXERCISES 9

A. Show, using Proposition 9.6, that every set of positive measure contains a non-measurable dense set.

B. A set B of real numbers is called a *Hamel basis* for the real numbers over the rationals if every real nonzero x can be written uniquely as

$$x = r_1 b_1 + \cdots + r_n b_n,$$

where $b_1, \ldots, b_n$ are distinct elements in B and $r_1, \ldots, r_n$ are nonzero rational numbers. (Using the axiom of choice, in particular, Zorn's lemma, we can prove that such a basis exists.) In the following we assume that there exists a Hamel basis.

Define $f: \mathbb{R} \to \mathbb{R}$ by

$$f(x) = r_1 + \cdots + r_n,$$

where $x = r_1 b_1 + \cdots + r_n b_n$ (uniquely written):
(a) show that $f(x + y) = f(x) + f(y)$; and
(b) prove that f is discontinuous everywhere.

C. Let $f: \mathbb{R} \to \mathbb{R}$ be such that $f(x) \neq 0$ for all $x \in \mathbb{R}$ and $f(x + y) = f(x)f(y)$ for all $x, y \in \mathbb{R}$. Suppose that f is bounded on a set with positive measure. Show that $f(x) = a^x$, where $a = f(1)$.

§10. Lebesgue Outer and Inner Measures

Up to now we have developed measure theory as a consequence of the theory of integration. In this section we prove that our method is consistent with that originally developed by Lebesgue. For this purpose we introduce the concepts of outer and inner measures.

10.1. Definition (Lebesgue, 1902). Given a set $A \subset [a, b]$, we define its *outer measure* $m^*(A)$ to be the infimum of $\sum_{n=1}^{\infty} |I_n|$ for all sequences of intervals $I_n \subset [a, b]$ such that $A \subset \bigcup_{n=1}^{\infty} I_n$.

It is immediately seen that a set of measure zero has the outer measure zero. We note that the domain of the outer measure is the family of *all* subsets of $[a, b]$.

The outer measure of a set A is approximated by those sets which are countable unions of intervals, but it is not necessary to consider the whole class of those sets. We have shown in Theorem 5.3, Chapter Zero, that every open set of real numbers is the union of a countable collection of mutually disjoint open intervals. In view of this result and the following lemma, it would be sufficient to take the open subsets of $[a, b]$ to approximate the outer measure of A.

10.2. Lemma. *If (I_n) is a sequence of intervals covering $A \subset [a, b]$, then for each $\varepsilon > 0$ there is an open set $G \subset [a, b]$ such that $G \supset A$ and*

$$m(G) < \sum_{n=1}^{\infty} |I_n| + \varepsilon.$$

Proof. Let a_n and b_n be the endpoints of the interval I_n. For each n, let

$$J_n = \left(a_n - \frac{\varepsilon}{2^{n+1}}, b_n + \frac{\varepsilon}{2^{n+1}} \right) \cap [a, b].$$

Set $G = \bigcup_{n=1}^{\infty} J_n$. Then G is open in $[a, b]$ and

$$m(G) \le \sum_{n=1}^{\infty} |J_n| < \sum_{n=1}^{\infty} |I_n| + \varepsilon. \qquad \square$$

10.3. Proposition. *Let A be a subset of $[a, b]$. Then*

$$m^*(A) = \inf\{m(G): G \supset A, G \text{ open}\}.$$

Proof. Let G be an open set in $[a, b]$ containing A. Then there is a sequence (J_n) of mutually disjoint inervals such that $G = \bigcup_{n=1}^{\infty} J_n$. Thus $m(G) = \sum_{n=1}^{\infty} |J_n|$. This shows that

$$m^*(A) \le m(G). \tag{1}$$

On the other hand, for every $\varepsilon > 0$ if (I_n) is a sequence of intervals covering A, then there is an open set $G \supset A$ such that

$$m(G) < \sum_{n=1}^{\infty} |I_n| + \varepsilon \tag{2}$$

by the preceding lemma. By relations (1) and (2), we obtain

$$m^*(A) \le \inf\{m(G): G \supset A, G \text{ open}\} \le m^*(A) + \varepsilon$$

from which the result follows, since $\varepsilon > 0$ is arbitrary. $\qquad \square$

10.4. Proposition. *Let A be a measurable subset of $[a, b]$. Then*

$$m^*(A) = m(A).$$

Proof. This is a consequence of Proposition 5.7. $\qquad \square$

The outer measure m^* is a function defined on the class of all subsets of $[a, b]$. It has the property that $m^*(A) = m(A)$ whenever A is measurable; in particular, $m^*(A)$ is the length of A if A is an interval. The value of m^* is nonnegative; m^* is monotonic in the sense that $m^*(A) \le m^*(B)$ if $A \subset B$ (see Exercise 10A). But it is not like the measure m. In fact, the outer measure m^* is not countably additive (see Exercise 10C). However, we have the following property:

10.5. Proposition. *Let (A_n) be a countable collection of subsets of $[a, b]$. Then*

$$m^*\left(\bigcup_{n=1}^{\infty} A_n\right) \leq \sum_{n=1}^{\infty} m^*(A_n).$$

Proof. Since each $m^*(A_n)$ is finite, then given $\varepsilon > 0$, there is an open set G_n containing A_n such that

$$m(G_n) < m^*(A_n) + \frac{\varepsilon}{2^n}.$$

Thus

$$m^*\left(\bigcup_{n=1}^{\infty} A_n\right) \leq m\left(\bigcup_{n=1}^{\infty} G_n\right) \leq \sum_{n=1}^{\infty} m(G_n)$$

$$< \sum_{n=1}^{\infty} \left[m^*(A_n) + \frac{\varepsilon}{2^n}\right] = \sum_{n=1}^{\infty} m^*(A_n) + \varepsilon.$$

Since ε was an arbitrary positive number,

$$m^*\left(\bigcup_{n=1}^{\infty} A_n\right) \leq \sum_{n=1}^{\infty} m^*(A_n). \qquad \square$$

Now we define the inner measure.

10.6. Definition. Let A be a subset of $[a, b]$. Then the *inner measure* $m_*(A)$ is defined as

$$m_*(A) = \sup\{m(F): F \subset A, F \text{ closed}\}.$$

Since $F \subset A$ implies $m(F) \leq m^*(A)$, we see that

$$m_*(A) \leq m^*(A).$$

10.7. Proposition.

$$m_*(A) = b - a - m^*([a, b]\backslash A).$$

Proof.

$$m_*(A) = \sup\{m(F): F \subset A, F \text{ closed}\}$$

$$= \sup\{b - a - m([a, b]\backslash F): F \subset A, F \text{ closed}\}$$

$$= b - a - \inf\{m([a, b]\backslash F): F \subset A, F \text{ closed}\}$$

$$= b - a - \inf\{m(G): G \supset [a, b]\backslash A, G \text{ open}\}$$

$$= b - a - m^*([a, b]\backslash A). \qquad \square$$

We restate Proposition 5.7 in terms of outer and inner measures.

10.8. Proposition. *Let A be a subset of $[a, b]$. Then A is measurable if and only if $m^*(A) = m_*(A)$. Furthermore, $m(A) = m^*(A) = m_*(A)$.*

10.9. Corollary. *Let A be a subset of $[a, b]$. Then A is measurable if and only if*

$$m^*(A) + m^*([a, b]\backslash A) = b - a.$$

Proof. If $m^*(A) = m_*(A)$, then

$$m^*(A) + m^*([a, b]\backslash A) = m_*(A) + m^*([a, b]\backslash A) = b - a.$$

Conversely, if $m^*(A) + m^*([a, b]\backslash A) = b - a$, then

$$m^*(A) = b - a - m^*([a, b]\backslash A) = m_*(A). \qquad \square$$

In the remaining part of this section we shall discuss the original method of Lebesgue [see Lebesgue (1926), Appendix in this book].

In his development of measure theory, Lebesgue called a set A *measurable* if $m_*(A) = m^*(A)$. Therefore, by Proposition 10.8, our definition, i.e., Riesz's definition, and Lebesgue's definition of a measurable set are essentially identical.

The second stage of Lebesgue's original method is marked by the concept of measurable function. In Lebesgue's sense, a function f is *measurable* if, for each constant c, the set

$$\{x: f(x) \leq c\}$$

is measurable. This is our Proposition 6.3.

Finally, Lebesgue proceeds to construct the definition of an integral for a bounded measurable function f. Let M be the upper bound and m the lower bound of f on $[a, b]$. We partition $[m, M]$ by $m = y_0 < y_1 < \cdots < y_n = M$. Let P denote this partition. Then we have the *Lebesgue sum* $L(f; P)$ defined by

$$L(f; P) = \sum_{j=0}^{n-1} y_j m\{x: y_j \leq f(x) < y_{j+1}\}.$$

It is this set $\{x: y_j \leq f(x) < y_{j+1}\}$ which plays the role analogous to the interval (x_j, x_{j+1}) in the Riemann sense of the integral, since it tells us the value of x which gives the $f(x)$ approximately equal values. The number $m\{x: y_j \leq f(x) < y_{j+1}\}$ is also meaningful, since f is measurable. The *Lebesgue integral* of f over $[a, b]$ is then defined as

$$\lim_{|P| \to 0} L(f; P).$$

This limit exists and is uniquely determined. (The limit $\lim_{|P| \to 0} L(f; P)$ is defined similarly to the definition of $\lim_{|P| \to 0} S(f; P)$ in §1, Chapter I.) The proof will be a part of the following discussion:

10.10. Proposition. *Let f be a bounded measurable function on $[a, b]$. Then $\lim_{|P| \to 0} L(f; P)$ exists and is unique. Furthermore,*

$$\int_a^b f(x)\, dx = \lim_{|P| \to 0} L(f; P),$$

where the left side is in the sense of Riesz.

Thus the integral defined by Lebesgue exists and is equal to the integral defined by Riesz.

Proof. We first define a *simple function* φ_P associated with the given partition P of $[m, M]$ by

$$\varphi_P(x) = y_j \qquad \text{if} \quad x \in \{x: y_j \le f(x) < y_{j+1}\}$$

for $j = 0, 1, \ldots, n - 1$. Then

$$\int_a^b \varphi_P(x)\, dx = L(f; P)$$

(in the sense of Riesz). Since

$$|f(x) - \varphi_P(x)| \le \max\{y_{j+1} - y_j : 0 \le j \le n\}$$

for any $\varepsilon > 0$ there is a partition P_ε with $|P_\varepsilon| < \varepsilon$ such that if P is a finer partition of $[m, M]$ than $P_{\varepsilon'}$, then $|f(x) - \varphi_P(x)| \le \varepsilon$. It follows that

$$\int_a^b f(x)\, dx = \lim_{|P| \to 0} \int_a^b \varphi_P(x)\, dx$$

by the Lebesgue Dominated Convergence Theorem. This shows that $\lim_{|P| \to 0} L(f; P)$ exists and is unique.   □

In the preceding proposition, we have shown that the *two definitions of the integral by Lebesgue and Riesz are equivalent for bounded measurable functions.* We shall now define the integral for unbounded functions as Lebesgue did. First, we define the integral for nonnegative unbounded measurable functions; then we extend this to general unbounded measurable functions.

If f is an unbounded and nonnegative measurable function, we truncate the function above by n for each natural number n; i.e., we consider the function

$$f_n = \min\{f, n\}.$$

Then f_n is integrable in both the Lebesgue and the Riesz sense, as we argued already. The sequence (f_n) is monotone increasing and converges to the function f. If

$$\lim \int_a^b f_n(x)\, dx$$

exists, then this limit is called the *Lebesgue integral* of f (in the sense of Lebesgue).

We show that this integral is equal to the integral defined by Riesz. Since f is the limit of a monotone increasing sequence of bounded integrable functions $f_n = \min\{f, n\}$, by the Monotone Convergence Theorem 5.5, Chapter II, f is integrable in the sense of Riesz and we have

$$\int_a^b f(x)\, dx = \lim \int_a^b f_n(x)\, dx.$$

Therefore, conversely, if f is integrable in the sense of Riesz and

$$M = \int_a^b f(x)\, dx,$$

then the truncated function f_n is bounded and integrable in either sense. Furthermore,

$$\int_a^b f_n(x)\, dx \le M \qquad \text{for each } n.$$

Therefore, by the Monotone Convergence Theorem, the limit $f(x) = \lim f_n(x)$ has the integral

$$\int_a^b f(x)\, dx = \lim \int_a^b f_n(x)\, dx.$$

Therefore f is integrable in the Lebesgue sense.

In the general case, if f is an unbounded measurable function, write $f = f^+ - f^-$. Then f is said to be *Lebesgue integrable* if both f^+ and f^- are integrable. Its integral is then defined by

$$\int_a^b f^+(x)\, dx - \int_a^b f^-(x)\, dx$$

and hence is equal to $\int_a^b f(x)\, dx$.

Hence the foregoing discussion gives the following proposition:

10.11. Proposition. *The theories of integration developed by Lebesgue and Riesz are equivalent; that is, they yield the same class of integrable functions with the same value for the integral of each.*

EXERCISES 10

A. Show that $m^*(A) \le m^*(B)$ if $A \subset B$.

B. If $m^*(A) = 0$, show that A is measurable.

C. Show that m^* is not countably additive.

D. Show that $m^*(A + x) = m^*(A)$.

E. Show that $m_*(A) = 0$, where A is the nonmeasurable set constructed in §4.

F. Let $A, B \subset [a, b]$ such that

$$\inf\{|x - y|: x \in A, y \in B\} > 0.$$

Then

$$m^*(A \cup B) = m^*(A) + m^*(B).$$

CHAPTER IV
Generalizations

We now undertake the task of generalizing the results of Chapters II and III, which relate to the case of a closed interval [a, b], to the case of more general sets. We could have made this generalization from the beginning, but it is the author's experience that small doses of abstraction step by step are pedagogically more sound than one full strength dose.

§1. The Integral on Measurable Sets

The concept of the Lebesgue integral which has been defined on a closed interval [a, b] may be defined on any measurable set E in [a, b].

1.1. Definition. Let E be a measurable set in [a, b]. A function $f: E \to \mathbb{R}^*$ is said to be *integrable* on E if the extended function $F: [a, b] \to \mathbb{R}^*$ defined by

$$F(x) = \begin{cases} f(x) & \text{if } x \in E, \\ 0 & \text{otherwise,} \end{cases}$$

is integrable on [a, b]. The *integral of f on E* is defined and denoted by

$$\int_E f(x) \, dx = \int_a^b F(x) \, dx.$$

A function $f: E \to \mathbb{R}^*$ is said to be *measurable* if the extended function F is measurable.

From these definitions, it is immediate that every *bounded measurable function on E is integrable* (see Proposition 1.5, Chapter III).

The integral defined on E satisfies the usual properties of the integral defined on $[a, b]$. We state several properties, some of which will be useful later. From §4, Chapter II, we have the following propositions. The proofs of these propositions follow readily from the definition of the integral on E.

1.2. Proposition. *If f and g are integrable on E and α and β are real numbers, then $\alpha f + \beta g$ is integrable on E and*

$$\int_E [\alpha f(x) + \beta g(x)]\, dx = \alpha \int_E f(x)\, dx + \beta \int_E g(x)\, dx.$$

That is, the space of all integrable functions on E, denoted by $L(E)$, is a vector space over the field $\mathbb{R}$.

1.3. Proposition. *If f and g are integrable on E and $f \le g$ almost everywhere on E, then*

$$\int_E f(x)\, dx \le \int_E g(x)\, dx.$$

1.4. Proposition. *If f is integrable on E, then $|f|$ is integrable on E and*

$$\left| \int_E f(x)\, dx \right| \le \int_E |f(x)|\, dx.$$

The Beppo Levi theorem holds for the integral on E. We formulate here the generalized Beppo Levi Theorem 5.7, Chapter II, for the integral on E.

1.5. The Beppo Levi Theorem. *Let $\sum_{n=1}^{\infty} f_n$ be a series of integrable functions on E such that*

$$\sum_{n=1}^{\infty} \int_E |f_n(x)|\, dx < \infty.$$

Then the series $\sum_{n=1}^{\infty} f_n$ converges almost everywhere on E, and if $f = \sum_{n=1}^{\infty} f_n$, then f is integrable on E and

$$\int_E f(x)\, dx = \sum_{n=1}^{\infty} \int_E f_n(x)\, dx.$$

We also generalize the Lebesgue Dominated Convergence Theorem.

1.6. The Lebesgue Theorem. *If a sequence (f_n) of integrable functions on E converges almost everywhere on E to a function f and if there is an integrable function g on E such that*

$$|f_n(x)| \le g(x)$$

for all n, then f is integrable on E and

$$\int_E f(x)\, dx = \lim \int_E f_n(x)\, dx.$$

We have stated above some properties of the integral on E which follow immediately from the definition of the integral on E. This section contains some more properties which would be expected of the Lebesgue integral if it is to be a useful concept.

1.7. Proposition (The First Mean Value Theorem). *If f is an integrable function on E and*

$$A \leq f(x) \leq B$$

for almost all x in E, then there exists a number C, $A \leq C \leq B$, such that

$$\int_E f(x)\, dx = Cm(E).$$

Proof.

$$Am(E) \leq \int_E f(x)\, dx \leq Bm(E). \qquad \square$$

1.8. Proposition. *If f is integrable on E, then f is integrable on each measurable subset of E. Furthermore, if $E = \bigcup_{n=1}^{\infty} E_n$ and the E_n's are mutually disjoint measurable sets, then*

$$\int_E f(x)\, dx = \sum_{n=1}^{\infty} \int_{E_n} f(x)\, dx$$

(countable additivity of the integral for measurable sets).

Proof. The first part is quite trivial (see Exercise 1C). We show the second part. Since the E_n's are mutually disjoint, we have

$$1 = \sum_{n=1}^{\infty} \chi_{E_n}$$

and

$$f = \sum_{n=1}^{\infty} \chi_{E_n} f$$

on E. It follows that the partial sums of this series are bounded above by the integrable function $|f|$. Therefore, by the Lebesgue Theorem 1.6, we can integrate the series term by term to obtain

$$\int_E f(x)\, dx = \sum_{n=1}^{\infty} \int_{E_n} f(x)\, dx. \qquad \square$$

The converse is not necessarily true; that is, it is not true in general that a function f is integrable on E if it is integrable on each E_n and the series

$$\sum_{n=1}^{\infty} \int_{E_n} f(x)\, dx$$

converges. The converse, however, with an additional hypothesis that $f \geq 0$ on each E_n, is true. More generally, we have the following proposition:

1.9. Proposition. *Let $E = \bigcup_{n=1}^{\infty} E_n$, where the E_n's are mutually disjoint measurable subsets of $[a, b]$. Let f be integrable on each E_n. Then f is integrable on E if and only if*

$$\sum_{n=1}^{\infty} \int_{E_n} |f(x)|\, dx < \infty.$$

If this is the case, we have

$$\int_E f(x)\, dx = \sum_{n=1}^{\infty} \int_{E_n} f(x)\, dx.$$

Proof. If f is integrable on E, then $|f|$ is also integrable on E. Therefore, by Proposition 1.8, we have

$$\sum_{n=1}^{\infty} \int_{E_n} |f(x)|\, dx < \infty,$$

and

$$\int_E f(\ )\, dx = \sum_{n=1}^{\infty} \int_{E_n} f(x)\, dx.$$

Conversely, suppose that

$$\sum_{n=1}^{\infty} \int_{E_n} |f(x)|\, dx < \infty.$$

Since $\int_{E_n} f(x)\, dx = \int_E \chi_{E_n}(x) f(x)\, dx$, applying the Beppo Levi Theorem 1.5 with $f_n = \chi_{E_n} f$, we obtain the result that f is integrable on E and

$$\int_E f(x)\, dx = \sum_{n=1}^{\infty} \int_{E_n} f(x)\, dx. \qquad \square$$

1.10. Corollary. *Let (E_n) be a monotone increasing sequence of measurable subsets of $[a, b]$; i.e.,*

$$E_1 \subset E_2 \subset \cdots.$$

If f is integrable on each E_n and

$$\lim \int_{E_n} |f(n)|\, dx < \infty,$$

then f is integrable on $E = \bigcup_{n=1}^{\infty} E_n$ and

$$\int_E f(x)\, dx = \lim \int_{E_n} f(x)\, dx.$$

Proof. Write

$$E = E_1 \cup (E_2 \backslash E_1) \cup (E_3 \backslash E_2) \cup \cdots$$

and apply the preceding proposition. $\square$

1.11. Corollary. *Suppose that $\varepsilon_1 > \varepsilon_2 > \cdots > \varepsilon_n > \cdots$, $\varepsilon_n \to 0$ as $n \to \infty$, and f is integrable on each $[a + \varepsilon_n, b]$. Then f is integrable on $[a, b]$ if and only if*

$$\lim \int_{a+\varepsilon_n}^b |f(x)|\, dx < \infty.$$

In this case, we have

$$\int_a^b f(x)\, dx = \lim \int_{a+\varepsilon_n}^b f(x)\, dx.$$

As a consequence of Corollary 1.11, *an arbitrary function f for which the Riemann integral*

$$\int_{a+\varepsilon}^b |f(x)|\, dx$$

approaches a finite limit as $\varepsilon \to 0$ is Lebesgue integrable on $[a, b]$, and the Lebesgue integral of f on $[a, b]$ is obtained by

$$\int_a^b f(x)\, dx = \lim_{\varepsilon \to 0} \int_{a+\varepsilon}^b f(x)\, dx.$$

It is quite unfortunate that we require the absolute integrability of f on $[a + \varepsilon, b]$. But it is a natural consequence of our procedure of defining integrability in terms of the class L^+ (or f^+ and f^-). These procedures rule out certain functions treated in *improper Riemann integration*. In this connection, we further remark that a function f which is improperly Riemann integrable on $[a, b]$ (i.e.,

$$\lim_{\varepsilon \to 0} \int_{a+\varepsilon}^b f(x)\, dx$$

exists) such that

$$\lim_{\varepsilon \to 0} \int_{a+\varepsilon}^b |f(x)|\, dx = \infty$$

cannot be Lebesgue integrable on $[a, b]$ (see Example 1.12).

It may be said that Lebesgue integration is absolute integration in the sense that f is Lebesgue integrable if and only if $|f|$ is Lebesgue integrable. In other words, *the Lebesgue integral integrates only those functions whose absolute value function is also integrable.*

1.12. Example. Let $f: [0, 1] \to \mathbb{R}$ be defined by

$$f(x) = \begin{cases} (1/x)\sin(1/x) & \text{if } x \neq 0, \\ 0 & \text{if } x = 0. \end{cases}$$

Then f is improperly Riemann integrable but not Lebesgue integrable on $[0, 1]$.

Proof. It is an exercise to show that $\int_\varepsilon^1 f(x)\, dx$ converges to a finite limit as $\varepsilon \to 0$. We now prove that $|f|$ is not Lebesgue integrable by showing that the convergence of the integral is conditional. For this, let $a_n = 1/(2n\pi + \frac{1}{2}\pi)$, $b_n = 1/(2n\pi + \frac{1}{4}\pi)$. Then

$$\int_{a_n}^{b_n} |f(x)|\, dx > \frac{1}{\sqrt{2}}\log\left(1 + \frac{1}{8n + 1}\right)$$

(Exercise 1D). Therefore,

$$\sum_{n=1}^m \int_{a_n}^{b_n} |f(x)|\, dx > \frac{1}{\sqrt{2}}\log\left[\prod_{n=1}^m \left(1 + \frac{1}{8n + 1}\right)\right]$$

$$> \frac{1}{\sqrt{2}}\log\left(\sum_{n=1}^m \frac{1}{8n + 1}\right) \to \infty, \qquad \text{as } m \to \infty.$$

Therefore, $|f|$ cannot be integrable on $[0, 1]$. $\qquad\qquad\qquad\qquad\qquad\qquad$ $\square$

In the above example, f is continuous on $[0, 1]$. A much simpler example of such a function which is not continuous on $[0, 1]$ is given in Exercise 1E.

It would be ideal if every improperly Riemann integrable function were Lebesgue integrable. But, unfortunately, this is not the case. Any modification of the definition to admit conditional convergence of integrals rather than absolute convergence will rule out the countable additivity of the integral for measurable sets described in Proposition 1.8 (see Exercise 1F).

It would be desirable to have the following three properties for a theory of integration on $[a, b]$:

(1) *Every bounded measurable function on $[a, b]$ is integrable.*
(2) *The integral satisfies countable additivity for measurable sets.*
(3) *Every improperly Riemann integrable function is integrable; that is, the theory should permit conditional convergence of integrals.*

The Lebesgue theory satisfies only the first two of these properties. There are other integrals satisfying properties (1) and (3). Such are Perron's integral and Denjoy's integral. We refer the interested reader to S. Saks, *Theory of the Integral*, Chapters 4 to 8. As remarked before, properties (2) and (3) are incompatible. Therefore, it is impossible for a theory of integration to have all three.

We close this section with the following important property of the Lebesgue integral, which is sometimes expressed by saying that the integral as a set function:

$$E \to \int_E f(x)\, dx$$

is absolutely continuous (see also §6, Chapter V).

1.13. Proposition. *If $|f|$ is integrable on $[a, b]$, then for each $\varepsilon > 0$ there corresponds a $\delta > 0$ such that if $E \subset [a, b]$ is measurable with $m(E) < \delta$, then*

$$\left| \int_E f(x)\, dx \right| < \varepsilon.$$

Proof. Since f is also integrable on $[a, b]$, for $\varepsilon > 0$, we can find a continuous function g on $[a, b]$ such that

$$\int_a^b ||f(x)| - g(x)|\, dx < \frac{\varepsilon}{2}$$

(see Proposition 7.4, Chapter II). Let $M = \sup |g(x)|$ on $[a, b]$, and let $\delta = \varepsilon/2M$. If $E \subset [a, b]$ is a measurable set with $m(E) < \delta$, then we have

$$\left| \int_E f(x)\, dx \right| \le \int_E |f(x)|\, dx$$

$$\le \int_E ||f(x)| - g(x)|\, dx + \int_E |g(x)|\, dx$$

$$< \frac{\varepsilon}{2} + \delta M = \frac{\varepsilon}{2} + \frac{\varepsilon}{2} = \varepsilon. \qquad \square$$

EXERCISES 1

A. Show that if $f(x) = c$ on E, then

$$\int_E f(x)\, dx = cm(E).$$

B. For the purpose of reviewing the Lebesgue integral developed in Chapter II, prove Propositions 1.2 to 1.6.

C. Show that if f is integrable on E, then f is integrable on each measurable subset of E.

D. Show that
$$\int_{a_n}^{b_n} \left| \frac{1}{x} \sin \frac{1}{x} \right| dx > \frac{1}{\sqrt{2}} \log\left(1 + \frac{1}{8n+1}\right),$$
where a_n, b_n are as in Example 1.12.

E. Let $f: [0, 1] \to \mathbb{R}$ be defined by
$$f(x) = \begin{cases} n(-1)^{n+1} & \text{if } 1/(n+1) < x \le 1/n, \\ 0 & \text{if } x = 0. \end{cases}$$

Show that f is not Lebesgue integrable on $[0, 1]$, but it is improperly Riemann integrable. The improper integral is equal to $1 - \log 2$.

F. Let f be as in Exercise E. Show that for any real number r there is a sequence (E_n) of mutually disjoint measurable sets (namely, intervals) in $[0, 1]$ such that $[0, 1] = \bigcup_{n=1}^{\infty} E_n$ and
$$\sum_{n=1}^{\infty} \int_{E_n} f(x)\, dx = r.$$

G. If f is an integrable function on E and
$$A \le f(x) \le B$$
for almost all $x \in E$, then for every integrable function g, $g \ge 0$, on E such that fg is integrable on E, show that
$$A \int_E g(x)\, dx \le \int_E f(x)g(x)\, dx \le B \int_E g(x)\, dx.$$

Take care not to apply this inequality when g is not positive!
Notice also that fg may not be integrable.

§2. The Integral on Infinite Intervals

Up until now our domain of integration has been the bounded interval $[a, b]$ and its measurable subsets. But the infinite intervals (a, ∞), $(-\infty, b)$, $(-\infty, \infty)$ present us with no difficulties; only slight modifications of the concepts of step functions and measurable functions are required.

We develop the theory of integration on the entire real line $\mathbb{R} = (-\infty, \infty)$ rather than on half-lines, since the integral on a half-line can be defined in the same way as integration on a measurable set, as studied in §1.

2.1. Definition. A function $\varphi: \mathbb{R} \to \mathbb{R}$ is said to be a *step function* if it takes constant values on a finite number of finite open intervals and zero elsewhere;

i.e.,

$$\varphi = \sum_{k=1}^{n} a_k \chi_{I_k},$$

where $I_k = (x_{k-1}, x_k)$ for $0 \le k \le n$ and $-\infty < x_0 < x_1 < \cdots < x_n < \infty$.

In this case we define the integral of φ by

$$\int_R \varphi(x)\, dx = \sum_{k=1}^{n} a_k (x_k - x_{k-1}).$$

Now we introduce the class L^+—as before—as the set of functions, each of which is the limit of a monotone increasing sequence (φ_n) of step functions, with

$$\int_R \varphi_n(x)\, dx \le A$$

for all n. The class L is then defined as the set of functions of the form $f - g$, where $f \in L^+$ and $g \in L^+$.

If $f \in L^+$ and $f = \lim \varphi_n$, where (φ_n) is a monotone increasing sequence of step functions with $\int_R \varphi_n(x)\, dx \le A$ for all n, we define the integral of f by

$$\int_R f(x)\, dx = \lim \int_R \varphi_n(x)\, dx.$$

If $f \in L$ and $f = g - h$, where $g \in L^+$ and $h \in L^+$, then we define the integral of f by

$$\int_R f(x)\, dx = \int_R g(x)\, dx - \int_R h(x)\, dx.$$

Most of the results in Chapter II stated for a bounded interval $[a, b]$ are also valid for $\mathbb{R}$ without modifications of their proofs. In the following, however, we will examine those proofs which do need some modification.

The proof of the First Fundamental Lemma 2.3, Chapter II, needs some modification since we used the compactness of $[a, b]$.

2.2. First Fundamental Lemma. *Let (φ_n) be a monotone decreasing sequence of nonnegative step functions. Then the sequence (φ_n) converges to 0 almost everywhere if and only if $\lim \int_R \varphi_n(x)\, dx = 0$.*

Proof. Since φ_1 is a step function, φ_1 vanishes on the complement of a compact interval $[a, b]$. Then for every $n \in \mathbb{N}$, φ_n vanishes on $[a, b]$ since (φ_n) is monotone decreasing. Hence (φ_n) can be considered as a sequence of step functions defined on $[a, b]$. The rest of the proof is identical to that of the first fundamental lemma in Chapter II.   □

Proposition 3.6, Chapter II, has no analogue for the entire line $\mathbb{R}$, since the Riemann integral is for a finite interval. All propositions except Proposition 1.7 in the preceding section hold for $E = \mathbb{R}$. In particular, Corollary 1.10 can be modified as follows.

2.3. Proposition. *Let (a_n) be a sequence of positive real numbers such that $a_n \uparrow \infty$. Further, suppose that $f \colon \mathbb{R} \to \mathbb{R}^*$ is integrable on each $[-a_n, a_n]$. Then f is integrable on $\mathbb{R}$ if and only if*

$$\lim \int_{-a_n}^{a_n} |f(x)|\, dx < \infty.$$

In this case, we have

$$\int_{\mathbb{R}} f(x)\, dx = \lim \int_{-a_n}^{a_n} f(x)\, dx.$$

Essentially, the preceding proposition states the following:

2.4. Corollary. *A function f is integrable on $\mathbb{R}$ if and only if f is integrable on each interval $[a, b]$ and there is an $M > 0$ such that*

$$\int_a^b |f(x)|\, dx < M$$

for all a and b.

2.5. Corollary. *If f is integrable on $\mathbb{R}$, then $|f|$ is integrable on $\mathbb{R}$. Also*

$$\left| \int_{\mathbb{R}} f(x)\, dx \right| \le \int_{\mathbb{R}} |f(x)|\, dx.$$

The fact that f is integrable on $\mathbb{R}$ implies $|f|$ is integrable on $\mathbb{R}$ makes the Lebesgue integral on $\mathbb{R}$ very different from the improper Riemann integral on $\mathbb{R}$. We should notice that the improper Riemann integral is not a special case of the Lebesgue integral on $\mathbb{R}$. Following the usage for infinite series, the Lebesgue integral is absolutely convergent; that is, it integrates only those functions whose absolute value function is also integrable.

Before closing this section we shall give an example of a function which is improper Riemann integrable but not Lebesgue integrable on $\mathbb{R}$.

2.6. Example. Let

$$f(x) = \begin{cases} \dfrac{\sin x}{x} & \text{if } x \ne 0, \\[2mm] 0 & \text{if } x = 0. \end{cases}$$

Then the improper Riemann integral

$$\int_0^\infty f(x)\, dx = \frac{\pi}{2}$$

exists [see, e.g., Spivak (1967), pp. 328–330]. However,

$$\int_0^\infty |f(x)|\, dx = \infty.$$

Thus, f is not Lebesgue integrable on ℝ.

EXERCISES 2

A. For the purpose of reviewing the Lebesgue integral developed in Chapter II, check all prooofs of statements with respect to $[a, b]$ appearing in §1 to §7, Chapter II, and modify them (if necessary) to the corresponding statements for the entire real line ℝ.

B. Give an example of a nonintegrable function f whose absolute value $|f|$ is integrable on ℝ. This shows that the converse of Corollary 2.4 may fail.

C. Which of the following functions are integrable on $[0, \infty)$?
   (1) The characteristic function of the rationals in $[0, \infty)$.
   (2) The characteristic function of the irrationals in $[0, \infty)$.

D. If f is nonnegative and improper Riemann integrable on ℝ, prove that f is Lebesgue integrable.

E. Show that if $x > 0$, then the function

$$t \to e^{-t} t^{x-1}$$

is integrable on $[0, \infty)$. Further, show that

$$\lim_{n \to \infty} \int_0^n \left(1 - \frac{t}{n}\right)^n t^{x-1}\, dt = \int_0^\infty e^{-t} t^{x-1}\, dt.$$

This function is known as the *gamma function*.

§3. Lebesgue Measure on ℝ

In §2, Chapter III, we introduced the concept of the Lebesgue measure on $[a, b]$. In the present section, we will obtain an obvious generalization of the Lebesgue measure to a family of sets in ℝ.

We know that any integrable function on the bounded interval $[a, b]$ is representable almost everywhere as a limit of a sequence of step functions (see Proposition 4.7, Chapter II). The proof of Proposition 4.7, Chapter II, without any modification, for the entire line ℝ, shows that every integrable function on ℝ is representable almost everywhere as a limit of a sequence of step functions on $\mathbb{R}$. However, the converse of this statement is not true; i.e., a function on $\mathbb{R}$ which is an almost everywhere limit of a sequence of step functions may not be integrable. For example, any nonzero constant function on $\mathbb{R}$ is not integrable on $\mathbb{R}$, although it is a limit of a sequence of step functions.

As in §1, Chapter III, we define a function $f\colon \mathbb{R} \to \mathbb{R}^*$ as *measurable* if there is a sequence (φ_n) of step functions on $\mathbb{R}$ which converges to f almost everywhere on $\mathbb{R}$.

From the Lebesgue theorem for $\mathbb{R}$ (see §2) we can conclude, quite easily, that if f is measurable on $\mathbb{R}$ and there is an integrable function g such that

$$|f(x)| \leq g(x),$$

then f is integrable on $\mathbb{R}$.

The sum, the product, the maximum, the minimum, and the quotient of two measurable functions are measurable; in the last case it is understood that the denominator is almost everywhere different from zero. As a consequence, the absolute value function of a measurable function is measurable (see Proposition 1.3, Chapter III).

The limit function of an almost everywhere convergent sequence of measurable functions is also measurable (see Proposition 1.6, Chapter III).

3.1. Proposition. *A function $f\colon \mathbb{R} \to \mathbb{R}^*$ is measurable on $\mathbb{R}$ if and only if f is measurable on each bounded interval $[a, b]$.*

Proof. It is clear that if f is measurable on $\mathbb{R}$, it is measurable on each bounded interval $[a, b]$.

Conversely, suppose that f is measurable on each bounded interval $[a, b]$. Then, in particular, f is measurable on $[-n, n]$ for each natural number n. Let

$$f_n(x) = \begin{cases} f(x) & \text{if } x \in [-n, n], \\ 0 & \text{otherwise.} \end{cases}$$

Then f_n is measurable on $\mathbb{R}$, and $f_n \to f$ as $n \to \infty$. But the limit function of a sequence of measurable functions is measurable. Therefore, f is measurable over $\mathbb{R}$. $\square$

3.2. Definition. Let $E \subset \mathbb{R}$. Then E is said to be *measurable* if the characteristic function χ_E is measurable. The measure $m(E)$ of the measurable set E is defined by the integral

$$m(E) = \int_{\mathbb{R}} \chi_E(x)\, dx$$

provided that χ_E is integrable. Otherwise, we define

$$m(E) = \infty.$$

According to this definition, every interval is measurable, with the measure of an interval being its length. In particular, if I is a bounded interval with endpoints a and b, where $a < b$, then $m(I) = b - a$. If I is unbounded, $m(I) = \infty$.

The following proposition follows easily from Proposition 3.1:

3.3. Proposition. *A set E is measurable if and only if for every bounded interval $[a, b]$, $E \cap [a, b]$ is measurable in $[a, b]$.*

It turns out that *the complement of a measurable set is measurable; the union of a sequence of measurable sets is measurable.*

3.4. Proposition. *The family $\mathcal{M}$ of all measurable sets in ℝ is a σ-algebra, and the measure m is countably additive on $\mathcal{M}$.*

Proof. The first part follows easily from Proposition 3.3 above and Proposition 2.3, Chapter III (see Exercise 3B). We now show that the measure m is countably additive. Let (E_n) be a sequence of mutually disjoint measurable sets. Let $E = \bigcup_{n=1}^{\infty} E_n$. If $\sum_{n=1}^{\infty} m(E_n) = \infty$, then it is trivial that

$$m(E) = \sum_{n=1}^{\infty} m(E_n).$$

If $\sum_{n=1}^{\infty} m(E_n) < \infty$, then consider $\chi_E = \sum_{n=1}^{\infty} \chi_{E_n}$. Then we have $f_k \uparrow \chi_E$, where $f_k = \sum_{n=1}^{k} \chi_{E_n}$. Hence, χ_E is integrable by the Lebesgue theorem and

$$m(E) = \int_{\mathbb{R}} \chi_E(x)\, dx = \sum_{n=1}^{\infty} \int_{\mathbb{R}} \chi_{E_n}(x)\, dx = \sum_{n=1}^{\infty} m(E_n). \qquad \square$$

3.5. Proposition. *If (E_n) is an increasing sequence of measurable sets in ℝ (i.e., $E_1 \subset E_2 \subset \cdots$), then*

$$m\left(\bigcup_{n=1}^{\infty} E_n \right) = \lim m(E_n).$$

Proof.

$$\bigcup_{n=1}^{\infty} E_n = E_1 \cup (E_2 \backslash E_1) \cup (E_3 \backslash E_2) \cup \cdots. \qquad \square$$

A result strictly analogous to Proposition 3.5 for the intersections of a decreasing sequence of measurable sets cannot be proven. To see this, let $E_n = [n, \infty)$. Then $m(E_n) = \infty$ for all $n = 1, 2, \ldots$, so that $\lim m(E_n) = \infty$. On the other hand,

$$m\left(\bigcap_{n=1}^{\infty} E_n \right) = m(\varnothing) = 0.$$

However, we can obtain the following generalization of Proposition 2.5(b), Chapter III:

3.6. Proposition. *If (E_n) is a decreasing sequence of measurable sets in $\mathbb{R}$ (that is, $E_1 \supset E_2 \supset \cdots$), and if $m(E_n)$ is finite for some n, then*

$$m\left(\bigcap_{n=1}^{\infty} E_n\right) = \lim m(E_n).$$

Proof. We may assume $m(E_1) < \infty$, so that the sequence $(E_1 \backslash E_n)$ is an increasing sequence. Then the result follows from the proof of Proposition 2.5, Chapter III.   □

The Egoroff theorem stated in §7, Chapter III, can be generalized to the set E on which the functions are defined to have finite measure (Exercise 3C). However, an analogue of the Egoroff theorem, due to Lusin, can be stated for arbitrary measurable subsets of $\mathbb{R}$.

3.7. Theorem (Lusin, 1912). *Let E be a measurable set in $\mathbb{R}$. Let (f_n) be a sequence of measurable functions defined on E, which converges to a function f almost everywhere on E. Then E can be written as the union*

$$E = A \cup \left(\bigcup_{n=1}^{\infty} B_n\right),$$

where $m(A) = 0$ and (f_n) converges to f uniformly on each B_n.

Proof. Let $E_n = E \cap ([-n, -n + 1) \cup (n, n + 1])$. Then

$$E = \bigcup_{n=1}^{\infty} E_n.$$

Therefore, E is the union of a countable number of mutually disjoint sets of finite measure. For each n, we apply the Egoroff theorem to the set E_n, and we can find a sequence of measurable subsets

$$E_{n_1}, E_{n_2}, \ldots$$

of E_n such that

$$m\left(E_n \backslash \bigcup_{j=1}^{k} E_{n_j}\right) < \frac{1}{k}$$

for all $k = 1, 2, \ldots$, and such that the sequence (f_n) converges to f uniformly on each E_{n_j}. Let $A_n = E_n \backslash \bigcup_{j=1}^{\infty} E_{n_j}$. Then $m(A_n) = 0$. We need only set $A = \bigcup_{n=1}^{\infty} A_n$ and rearrange the double sequence (E_{n_j}) to a single sequence (B_n) to obtain the desired result.   □

If f is measurable on $\mathbb{R}$, we define a set function μ on all measurable subsets of $\mathbb{R}$ by

$$\mu(E) = \begin{cases} \iint_E |f(x)|\, dx & \text{if } f \text{ is integrable on } E, \\ \infty & \text{otherwise.} \end{cases}$$

Then:

(1) $0 \leq \mu(E) \leq \infty$ for all measurable subsets of $\mathbb{R}$.
(2) $\mu(\emptyset) = 0$.
(3) μ is countably additive, in the sense that if (E_n) is a mutually disjoint sequence of measurable sets in $\mathbb{R}$, then

$$\mu\left(\bigcup_{n=1}^{\infty} E_n\right) = \sum_{n=1}^{\infty} \mu(E_n).$$

If we take $f = 1$, then we have the Lebesgue measure $m(E) = \mu(E)$.

Thus μ generalizes the Lebesgue measure.

In an advanced course of measure theory we usually study a more general measure. In general, a *measure* is a set function μ defined on a σ-algebra $\mathscr{A}$ of subsets of a given set X such that:

(1) $0 \leq \mu(E) \leq \infty$ for all $E \in \mathscr{A}$;
(2) $\mu(\emptyset) = 0$; and
(3) μ is countably additive.

An element of the σ-algebra $\mathscr{A}$ is said to be μ-measurable.

One can also define integration on X in the fashion of Lebesgue (see §10, Chapter III). We refrain from a deeper discussion of such a theory here, because such study was not our original intention. We refer the interested reader to Halmos (1950).

EXERCISES 3

A. Prove Proposition 3.1 in detail.

B. Show that the family $\mathscr{M}$ of all measurable sets in $\mathbb{R}$ is a σ-algebra.

C. **The Egoroff Theorem.** *Let E be a measurable set in $\mathbb{R}$ such that $m(E) < \infty$. Let (f_n) be a sequence of measurable functions on E which converges to a function f almost everywhere on E. Then (f_n) converges to f almost uniformly.*

D. Let X be a set, and let $\mathscr{A}$ be a σ-algebra of subsets of X. Define on X the following:
   (1) a measurable function;
   (2) an integrable function; and
   (3) the integral of the preceding function.

§4. Finite Additive Measure: The Banach Measure Problem

In summary, we see that the Lebesgue measure on $\mathbb{R}$ is a function $m: \mathscr{M} \rightarrow \mathbb{R}^+ \cup \{\infty\}$, where $\mathscr{M}$ is the family of all measurable sets in $\mathbb{R}$, satisfying the following conditions:

(i) $m([0, 1]) = 1$;
(ii) m is countably additive; that is, if (E_n) is a sequence of mutually disjoint sets in $\mathcal{M}$,

$$m\left(\bigcup_{n=1}^{\infty} E_n\right) = \sum_{n=1}^{\infty} m(E_n); \quad \text{and}$$

(iii) m is translation invariant; that is, if $E \in \mathcal{M}$ and $x \in \mathbb{R}$, then $E + x \in \mathcal{M}$ and $m(E + x) = m(E)$.

Unfortunately, as we saw in §4, Chapter III, there are sets which are nonmeasurable. This suggests that the Lebesgue measure is not as nice as we might want, if we assume the axiom of choice. Therefore, it is natural to ask whether it is possible to improve somewhat on Lebesgue's definition.

Ideally, we would like to find a set function satisfying the following conditions:

(1) $\mu(E)$ is defined for all $E \subset \mathbb{R}$;
(2) $\mu([0, 1]) = 1$;
(3) μ is countably additive; and
(4) μ is translation invariant.

As we have seen before in §4, Chapter III, it is impossible to find a set function satisfying all four of these conditions, since the last two conditions—taken together—contradict the first condition. Therefore, we must sacrifice one of these conditions.

The existence of a set function satisfying the first three conditions is not known. The Lebesgue measure is a set function m satisfying the last three conditions. It should be remarked that the Lebesgue measure is not the only possible measure satisfying conditions (2), (3), and (4). S. Kakutani and J.C. Oxtoby have constructed an extension of Lebesgue measure in "A non-separable translation invariant extension of the Lebesgue measure space" (1950). The family of Kakutani–Oxtoby measurable sets is enormously larger than that of Lebesgue measureable sets.

It is now natural to ask what parts of the last three conditions (2) to (4) we must sacrifice in order to retain the first condition. Since the concept of a measure should generalize the length of intervals, the requirement $\mu([0, 1]) = 1$ is legitimate. We would also like to retain condition (4) because we hope that two "congruent" sets would have the same measure.

Under these considerations, it is now clear that we should either replace or weaken condition (3) by some other properties. There are two approaches. The first alternative is *countable subadditivity*; that is, if (E_n) is a sequence of mutually disjoint sets of real numbers, then

$$\mu\left(\bigcup_{n=1}^{\infty} E_n\right) \leq \sum_{n=1}^{\infty} \mu(E_n).$$

The countable subadditivity is satisfied by the Lebesgue outer measure (see Proposition 10.4, Chapter III). But the outer measure is not by itself of great use.

Another possible alternative to condition (3) is *finite additivity*; that is, for every finite, mutually disjoint class $E_1, \ldots, E_n$ of sets in $\mathbb{R}$, we have

$$\mu\left(\bigcup_{k=1}^{\infty} E_k\right) = \sum_{k=1}^{n} \mu(E_k)$$

Stephan Banach, in his "Sur le problème de mesure" (1923), produced the following theorem:

4.1. Theorem (Banach, 1923). *There exists a set function μ defined for all subsets of $\mathbb{R}$ such that*:

(i) $0 \leq \mu(E) \leq \infty$ *for all* $E \subset \mathbb{R}$;
(ii) $\mu([0, 1]) = 1$;
(iii) μ *is finitely additive; and*
(iv) μ *is translation-invariant.*

The proof of Banach's theorem requires acquaintance with functional analysis, in particular the Hahn–Banach theorem, and is beyond the scope of this book. For a readable proof, see E. Hewitt, *Theory of Functions of a Real Variable* (1960), pp. 99–109. We also remark that such a function μ is not unique.

It is an interesting fact that an integral can be defined by a set function as described in Theorem 4.1, but there is no analogue of the limit theorems (the Beppo Levi theorem; the Lebesgue theorem) because of finite additivity.

These remarks are sufficient to point out the superiority of the Lebesgue theory.

In connection with the Banach theorem, we indicate the existence problem of such a function for the n-dimensional Euclidean space $\mathbb{R}^n$. We first formulate the problem.

4.2. Banach Measure Problem for $\mathbb{R}^n$. *Construct a set function μ defined for all subsets of $\mathbb{R}^n$ such that*:

(i) $0 \leq \mu(E) \leq \infty$ *for all* $E \subset \mathbb{R}^n$;
(ii) $\mu(I^n) = 1$, *where* $I^n = [0, 1] \times \cdots \times [0, 1]$, n *times*;
(iii) μ *is finitely additive; and*
(iv) $\mu(A) = \mu(B)$ *if A and B are isometric.*

(Two sets A and B are called *isometric* or *congruent* if there is a one–one correspondence f between A and B such that $\|f(x) - f(y)\| = \|x - y\|$.)

Banach studied this problem for $n = 1, 2$ and solved it affirmatively (see Theorem 4.1).

The German mathematician Felix Hausdorff (1968–1942) solved this problem negatively for $n \geq 3$. Banach and Hausdorff's results are summarized in the following theorem:

4.3. Theorem (Banach, Hausdorff). *The Banach measure problem for $\mathbb{R}^n$ is solvable if and only if $n = 1, 2$.*

The proof of Hausdorff's result can be found in *Grundzüge der Mengenlehre* (1914), pp. 469–472.

The unsolvability of the finite measure problem for $\mathbb{R}^3$ is derived from the following lemma, which is known as *Hausdorff's paradox*:

4.4. Lemma. *The unit sphere S^2 in $\mathbb{R}^3$ can be decomposed into four mutually disjoint sets*

$$S^2 = A \cup B \cup C \cup D,$$

where D is a countable set and $A \equiv B \equiv C \equiv B \cup C$ ($\equiv$ denotes congruence between sets).

For a proof of this remarkable lemma, we refer the interested reader to Natanson, *Theory of Functions of a Real Variable* (1960), Appendix V, or to Sierpinski's monograph, "On the congruence of sets and their equivalence by finite decomposition" (1953).

Using Lemma 4.4, we can easily show that the Banach measure problem for $\mathbb{R}^3$ is unsolvable. In fact, if there is a set function μ which is finitely additive, then

$$\mu(S^2) = \mu(A) + \mu(B) + \mu(C) + \mu(D) = 3\mu(A) + \mu(D),$$

and

$$\mu(S^2) = \mu(A) + \mu(B \cup C) + \mu(D) = 2\mu(A) + \mu(D),$$

which is an obvious contradiction. Therefore the problem is unsolvable. For $n > 3$, we leave the proof to the reader (Exercise 4B).

Related to Hausdorff's paradox is the *Banach–Tarski paradox*, named after two distinguished Polish mathematicians, Stephan Banach (1892–1945) and Alfred Tarski (1902–   ), "Sur la décomposition des ensembles de points en parties respectivement congruentes" (1924). This paradoxical theorem of Banach and Tarski is so astounding and unbelievable that its equal may be found nowhere in the annals of mathematics. Roughly, the theorem states that a sphere can be decomposed into finitely many disjoint parts which can be reassembled, after suitable rotations and translations, to give surfaces of two spheres, each of the original radius. The proof of this theorem uses Lemma 4.4. The conclusion, though obtained by rigorous and impeccable arguments, seems almost as incredible to the mathematician as to the layman [see also Robinson (1947), Stromberg (1979), and Dekker and de Groot (1956)].

EXERCISES 4

A. **Jordan Content.** Let A be a bounded subset of $\mathbb{R}$. The *content* of A was defined by the French mathematician Camille Jordan (1838–1922) to be the number

$$C(A) = \inf\left\{\sum_{k=1}^{n} (b_k - a_k) : A \subset \bigcup_{k=1}^{n} (a_k, b_k)\right\},$$

where the (a_k, b_k) are mutually disjoint. Clearly, this is always a well-defined number; however, it has proved to be unsatisfactory for the purposes of modern analysis. By making an apparently slight—but actually vital—change in the above definition, Henri Lebesgue produced the Lebesgue measure. Lebesgue's fundamental idea was to consider countable coverings of A, while Jordan considered only finite coverings of the set in question.

(a) Show that the Jordan content of the set of all rational numbers in [0, 1] is equal to 1.

(b) Show that the Jordan content of the set B of all irrational numbers in [0, 1] is equal to 1.

(c) Show that the Jordan content is not finitely additive.

B. Show that the Banach measure problem for $\mathbb{R}^n$, $n > 3$ is unsolvable. *Hint*: If this is solvable for $\mathbb{R}^{n+1}$, it is also solvable for $\mathbb{R}^n$.

C. Let μ be a set function which is not identically ∞. Suppose that μ is countably additive and translation invariant. Show that $\mu(A) = 0$ for any countable set A.

D. Show that there is no $\{0, 1\}$-valued measure μ defined on the family of all subsets of $\mathbb{R}$ such that $\mu(\mathbb{R}) = 1$ and $\mu(\{x\}) = 0$ for all $x \in \mathbb{R}$. For related problems, see Gillman and Jerison (1960), Chap. 12.

§5. The Double Lebesgue Integral and the Fubini Theorem

For a function of several variables we may define the Lebesgue integral by exactly the same process as for a function of one variable. We will consider only functions of two variables, since this will make the general case clear. In this special case, rectangles and squares will play the role of intervals, and we will need only to imitate the definitions and methods which we used for functions of one variable.

We restrict ourselves to the case of functions defined on the rectangle $S = [a, b] \times [c, d]$. If we have developed the Lebesgue theory on S, then it is easy to generalize the theory to a measurable set in S; similarly, we can develop the theory for the entire plane $\mathbb{R}^2$ analogously to that for the real line $\mathbb{R}$.

In this section S will represent the rectangle $[a, b] \times [c, d]$.

5.1. Definition. A subset A of $\mathbb{R}^2$ is said to be *a set of measure zero* if for any $\varepsilon > 0$ there exists a sequence (R_n) of rectangles such that $A \subset \bigcup_{n=1}^{\infty} R_n$ and $\sum_{n=1}^{\infty} |R_n| < \varepsilon$, where $|R_n|$ denotes the area of the rectangle R_n.

Thus any subset of the real axis in $\mathbb{R}^2$ is a set of measure zero. It is easy to see that if A and B are sets of measure zero in $\mathbb{R}$, then the product $A \times B$ is a set of measure zero. More generally, if A is a set of measure zero in $\mathbb{R}$, then for any subset B of $\mathbb{R}$, the product set $A \times B$ is a set of measure zero (see

Exercise 5A). A property which holds for all points of S outside of some set of measure zero is said to hold *almost everywhere* (a.e.) in S.

5.2. Definition. If the rectangle $S = [a, b] \times [c, d]$ is decomposed into a finite number of rectangles $R_1, R_2, \ldots, R_n$, then a function which assumes a constant value on the interior of each of these rectangles is called a *step function*. We can disregard the values of a step function on the edges of the rectangles or assign values to the function arbitrarily there. Such a step function is denoted by

$$\varphi(x, y) = \sum_{k=1}^{n} a_k \chi_{r_k}(x, y),$$

where r_k denotes the interior of R_k.

5.3. Definition. We define the *integral of a step function* φ, in a natural way, by the formula

$$\iint_S \varphi(x, y)\, dx\, dy = \sum_{k=1}^{n} a_k |R_k|.$$

If $\varphi(x, y) \geq 0$ for all x, y in S, then the integral is the volume of the solid between the surface $\{(x, y, \varphi(x, y)): (x, y) \in S\}$ and the xy plane (see Figure 4.1). Since the rectangle S can be decomposed into a finite number of subrectangles in many different ways, we must check if $\iint_S \varphi(x, y)\, dx\, dy$ is defined uniquely. In fact, if $\varphi(x, y) = \sum_{k=1}^{m} a_k \chi_{r_k} = \sum_{k=1}^{n} b_k \chi_{s_k}$, where $\{R_1, \ldots, R_m\}$ and $\{S_1, \ldots, S_n\}$ are two rectangular decompositions of S, and r_k and s_k are interiors of R_k and S_k, respectively, then we can obtain a common

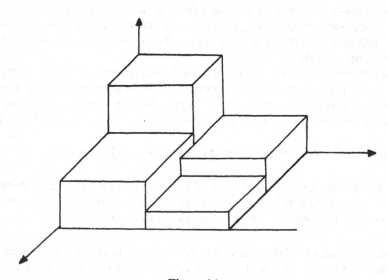

Figure 4.1

refinement $\{T_1, \ldots, T_t\}$ of these two decompositions in terms of disjoint rect-angles such that $\varphi = \sum_{k=1}^{t} c_k \chi_{t_k}$ almost everywhere for some suitably chosen c_k's. It is easy to check that both $\sum_{k=1}^{m} a_k |R_k|$ and $\sum_{k=1}^{n} b_k |S_k|$ are equal to $\sum_{k=1}^{t} c_k |T_k|$. Therefore, the integral $\iint_S \varphi(x, y)\, dx\, dy$ does not depend on a particular choice of a rectangular decomposition of S.

The proofs of the following results are entirely analogous to those in Chapter II, with intervals replaced by rectangles:

5.4. First Fundamental Lemma. *Let (φ_n) be a monotone decreasing sequence of nonnegative step functions defined on S. Then $\varphi_n \downarrow 0$ almost everywhere on S if and only if $\lim \iint_S \varphi_n(x, y)\, dx\, dy = 0$.*

5.5. Second Fundamental Lemma. *If (φ_n) is a monotone increasing sequence of step functions defined on S for which the sequence $(\iint_S \varphi_n(x, y)\, dx\, dy)$ converges, then the sequence (φ_n) converges almost everywhere on S.*

The converse of the Second Fundamental Lemma 5.5, as before in Propo-sition 2.6, Chapter II, is also true.

5.6. Proposition. *A set $A \subset S$ is of measure zero if and only if there exists a monotone increasing sequence (φ_n) of step functions on S such that $(\iint_S \varphi_n(x, y)\, dx\, dy)$ converges and (φ_n) diverges on A.*

We introduce the classes L^+ and L as before.

5.7. Definition. The class L^+ is the set of all measurable functions f, each of which is a limit of a monotone increasing sequence (φ_n) of step functions such that

$$\iint_S \varphi_n(x, y)\, dx\, dy \le A \qquad \text{for all } n,$$

where A is a constant. Then we define the *integral* of f by

$$\iint_S f(x, y)\, dx\, dy = \lim \iint_S \varphi_n(x, y)\, dx\, dy.$$

The question of whether this integral is well defined depends as before on both fundamental lemmas.

Finally, we define the *Lebesgue space L* as the set of differences $f = f_1 - f_2$, where $f_i \in L^+$. The integral of f, then, is defined by

$$\iint_S f(x, y)\, dx\, dy = \iint_S f_1(x, y)\, dx\, dy - \iint_S f_2(x, y)\, dx\, dy.$$

We also define a measurable function on S is a natural way: A function f is said to be *measurable* on S if it is representable as a limit of a sequence (φ_n) of step functions which converges almost everywhere on S.

All propositions in Chapter II stated for a Lebesgue integrable function on a closed interval $[a, b]$ are easily carried over to the corresponding propositions for the double integral on the rectangle S without modifying the actual proofs in Chapter II, except for replacing $f(x)$ by $f(x, y)$.

However, an important new problem arises here. The following question and its answer are of fundamental significance in many problems in analysis: If f is integrable on the rectangle $S = [a, b] \times [c, d]$, is the value of the integral $\iint_S f(x, y) \, dx \, dy$ equal to that of the repeated integral $\int_c^d [\int_a^b (x, y) \, dx] \, dy$ and the repeated integral $\int_a^b [\int_c^d f(x, y) \, dy] \, dx$? In classical analysis, this is true for continuous functions on S (see your calculus text). But, it is far from obvious that the existence of the integral $\iint_S f(x, y) \, dx \, dy$ guarantees the existence of either repeated integral. Historically, this question had already been considered for continuous functions by A.L. Cauchy in the early nineteenth century. It was finally solved by the Italian mathematician Guido Fubini (1879–1943) in 1907. See G. Fubini, "Sugli integrali multipli" (1907). Beppo Levi stated the result without proof in a footnote on p. 322 in his paper, "Sul principio di Dirichlet" (1906b). Also Lebesgue (1904) considered this problem for bounded integrable functions before Fubini.

The following example may motivate the reader to consider the Fubini theorem and may also serve to indicate the difficulties which must be overcome in establishing the theorem.

5.8. Example. Let I be the unit square $[0, 1] \times [0, 1]$. Let $A \subset [0, 1]$ be a nonmeasurable set, and let $\bar{A} = A \times \{0\}$. Then $\bar{A} \subset I$, and $\bar{A}$ is a set of measure zero in $\mathbb{R}^2$ (see Exercise 5A). Therefore, the characteristic function of $\bar{A}$ is integrable on I, and its integral is zero. Denote $f = \chi_{\bar{A}}$. Then

$$\iint_I f(x, y) \, dx \, dy = 0.$$

Also, for every x in $[0, 1]$,

$$\int_0^1 f(x, y) \, dy = 0$$

and hence

$$\int_0^1 \left[\int_0^1 f(x, y) \, dy \right] dx = 0.$$

On the other hand, for every $y \neq 0$,

$$\int_0^1 f(x, y) \, dx = 0.$$

If $y = 0$, then

$$\int_0^1 f(x, y) \, dx = \int_0^1 \chi_A(x) \, dx$$

does not exist, since A is nonmeasurable. But the function

$$y \rightarrow \int_0^1 f(x, y) \, dx$$

is almost everywhere zero. Therefore, it is integrable on $[0, 1]$ and its integral

$$\int_0^1 \left[\int_0^1 f(x, y) \, dx \right] dy = 0$$

or

$$\iint_I f(x, y) \, dx \, dy = \int_0^1 \left[\int_0^1 f(x, y) \, dx \right] dy. \qquad \square$$

We now state the main theorem of this section.

5.9. The Fubini Theorem. *Let f be integrable on the rectangle $S = [a, b] \times [c, d]$. Then:*

(1) *For almost all y the function $x \rightarrow f(x, y)$ is integrable on $[a, b]$.*
(1′) *For almost all x the function $y \rightarrow f(x, y)$ is integrable on $[c, d]$.*
(2) *The function $y \rightarrow \int_a^b f(x, y) \, dx$ is integrable on $[c, d]$.*
(2′) *The function $x \rightarrow \int_c^d f(x, y) \, dy$ is integrable on $[a, b]$.*
(3) *The following equalities hold:*

$$\iint_S f(x, y) \, dx \, dy = \int_c^d \left[\int_a^b f(x, y) \, dx \right] dy = \int_a^b \left[\int_c^d f(x, y) \, dy \right] dx.$$

Because of the symmetry between x and y, it is sufficient to prove (1), (2), and the first equality of (3). It is easy to see that the Fubini theorem is valid for step functions (see Exercise 5B). If the conclusion of the theorem holds for each of two functions in L^+, it also holds for their differences, and hence it suffices to consider the case when f is in the class L^+. To show that the Fubini theorem holds true for functions of the class L^+, we need the following lemma. To avoid unnecessary confusion, a set of measure zero in $\mathbb{R}$ defined in Chapter I will be called a set of *linear measure zero*.

5.10. Fubini Lemma. *Let $A \subset S$ be a set of measure zero and let $Ay = \{x: (x, y) \in A\}$ be the y section of A. Then Ay is of linear measure zero for almost all y in $[c, d]$.*

Proof. By Proposition 5.6, there is a monotone increasing sequence (φ_n) of step functions on S for which $(\iint_S \varphi_n(x, y) \, dx \, dy)$ converges but $(\varphi_n(x, y))$ diverges for all (x, y) in A. Since the Fubini theorem is valid for step functions, the step functions defined by

$$\psi_n(y) = \int_a^b \varphi_n(x, y) \, dx$$

satisfy

$$\int_c^d \psi_n(y)\, dy = \iint_S \varphi_n(x, y)\, dx\, dy.$$

Since $\psi_n \uparrow$ and $(\int_c^d \psi_n(y)\, dy)$ converges, $(\psi_n(y))$ converges to a function $\psi(y)$ for almost all y in $[c, d]$ by the second fundamental lemma in Chapter II. Let $y' \in [c, d]$ be such that $\psi_n(y') \to \psi(y')$; i.e., $(\int_a^b \varphi_n(x, y')\, dx)$ converges. Again by the second fundamental lemma in Chapter II, $(\varphi_n(x, y'))$ converges for almost all $x \in [a, b]$. But if $x \in Ay'$, $(\varphi_n(x, y'))$ diverges, and hence Ay' is a set of linear measure zero.   □

Proof of Theorem. Let (φ_n) be a monotone increasing sequence of step functions which converges almost everywhere to f. For each n, we define a function ψ_n on $[c, d]$ by

$$\psi_n(y) = \int_a^b \varphi_n(x, y)\, dx.$$

Then $\psi_n(y)$ is defined for almost all y in $[c, d]$ ($\psi_n(y)$ cannot be defined if y is a point of $[c, d]$ such that the horizontal line passing through the point $(0, y)$ contains an edge of a rectangle R_k in the definition of the step function φ_n). The sequence (ψ_n) is then monotone increasing and the integral of ψ_n is bounded above for all n. In fact,

$$\int_c^d \psi_n(y)\, dy = \int_c^d \left[\int_a^b \varphi_n(x, y)\, dx \right] dy = \iint_S \varphi_n(x, y)\, dx\, dy$$

$$\leq \iint_S f(x, y)\, dx\, dy,$$

since φ_n is a step function and for this the Fubini theorem is valid. By the Monotone Convergence Theorem 5.5, Chapter II, we conclude that (ψ_n) converges almost everywhere on $[c, d]$ to an integrable function ψ and

$$\int_c^d \psi(y)\, dy = \lim \int_c^d \left[\int_a^b \varphi_n(x, y)\, dx \right] dy$$

$$= \lim \iint_S \varphi_n(x, y)\, dx\, dy = \iint_S f(x, y)\, dx\, dy. \qquad (1)$$

Let A be a subset of S on which (φ_n) fails to converge to f. Then A is of measure zero. It follows from Lemma 5.10 that $Ay = \{x : (x, y) \in A\}$ is of linear measure zero for almost all y in $[c, d]$. Let $y' \in [c, d]$ be such that Ay' is of linear measure zero and $\psi_n(y') \to \psi(y')$. Then $(\int_a^b \varphi_n(x, y')\, dx)$ converges, and hence, by the Second Fundamental Lemma 2.4, Chapter II, $\varphi_n(x, y') \to f(x, y')$ almost everywhere on $[a, b]$ and $x \to f(x, y')$ is integrable on $[a, b]$. Now

$$\int_a^b f(x, y')\, dx = \lim \int_a^b \varphi_n(x, y')\, dx = \psi(y') \qquad (2)$$

by the Monotone Convergence Theorem 5.3, Chapter II. The last relation (2) holds for almost all y' in $[c, d]$. Thus, we have from relations (1) and (2),

$$\iint_S f(x, y) \, dx \, dy = \int_c^d \psi(y) \, dy = \int_c^d \left[\int_a^b f(x, y) \, dx \right] dy. \qquad \square$$

According to the Fubini theorem, a double integral $\iint_S f(x, y) \, dx \, dy$ is computed by integrating first with respect to x and then with respect to y, or vice versa.

It should be noticed that the existence and equality of two repeated integrals does not guarantee the existence of the double integral (see Exercise 5F).

The following extension of the Fubini theorem is due to the English mathematician, E.W. Hobson (1856–1933), "On some fundamental properties of Lebesgue integrals in a two-dimensional domain" (1909). Also, the Italian mathematician L. Tonelli discovered this theorem independently in "Sull'integrazione per parti" (1909).

5.11. The Fubini–Hobson–Tonelli Theorem. *Let f be a measurable function on $S = [a, b] \times [c, d]$. Then, if either of the repeated integrals*

$$\int_a^b \left[\int_c^d |f(x, y)| \, dy \right] dx$$

or

$$\int_c^d \left[\int_a^b |f(x, y)| \, dx \right] dy$$

exists, f is integrable on S and hence

$$\iint_S f(x, y) \, dx \, dy = \int_a^b \left[\int_c^d f(x, y) \, dy \right] dx = \int_c^d \left[\int_a^b f(x, y) \, dx \right] dy.$$

We use the following fact without proof: Let f be a bounded and measurable function on S. Then f is integrable on S (see Proposition 1.5, Chapter III). We also use the Monotone Convergence Theorem for double integrals.

Proof. Suppose that the repeated integral

$$\int_c^d \left[\int_a^b |f(x, y)| \, dx \right] dy = A$$

exists. This means that

(a) for almost all y in $[c, d]$, the function $x \to |f(x, y)|$ is integrable on $[a, b]$, and
(b) the function $y \to \int_a^b |f(x, y)| \, dx$, which is defined for almost all y by (a), is integrable on $[c, d]$.

Let $f_n(x, y)$ be the function $|f(x, y)|$ truncated above by n; i.e.,

$$f_n(x, y) = \min\{|f(x, y)|, n\}.$$

Then f_n is bounded and measurable on S, and hence is integrable on S. By the

Fubini Theorem, we have

$$\iint_S f_n(x, y) \, dx \, dy = \int_c^d \left[\int_a^b f_n(x, y) \, dx \right] dy \leq A.$$

Therefore the sequence (f_n) satisfies the Monotone Convergence Theorem for double integrals, and $|f(x, y)|$ is integrable on S. Since f is measurable, we can conclude that f is integrable on S. But then we can apply the Fubini theorem and obtain the desired identity.   □

5.12. Corollary. *If f is measurable and nonnegative on $S = [a, b] \times [c, d]$, and one of the repeated integrals exists, then f is integrable on S and*

$$\iint_S f(x, y) \, dx \, dy = \int_a^b \left[\int_c^d f(x, y) \, dy \right] dx = \int_c^d \left[\int_a^b f(x, y) \, dx \right] dy.$$

Exercise 5D shows that we cannot omit the hypotheses of nonnegativity from the preceding corollary and absolute integrability from the Fubini–Hobson–Tonelli theorem.

Both the Fubini theorem and the Fubini–Hobson–Tonelli theorem hold for integrals over all of $\mathbb{R}^2$. Indeed, all the above theory of integration on a rectangle S may be extended easily to integrals on all of $\mathbb{R}^2$ or to the integrals on any measurable subsets of $\mathbb{R}^2$.

As an application of the Fubini–Hobson–Tonelli theorem, we have the following useful property of the double integral.

5.13. Proposition. *If f and g are integrable on $[a, b]$, then the function $(x, y) \to f(x)g(y)$ is integrable on $S = [a, b] \times [a, b]$ and*

$$\iint_S f(x)g(y) \, dx \, dy = \left[\int_a^b f(x) \, dx \right]\left[\int_a^b g(y) \, dy \right].$$

Proof. It is easy to show that the function $(x, y) \to f(x)g(y)$ is measurable (Exercise 5G) and that

$$\int_a^b \left[\int_a^b |f(x)| \, |g(y)| \, dx \right] dy = \left[\int_a^b |g(y)| \, dy \right]\left[\int_a^b |f(x)| \, dx \right]$$

exists. Therefore, the said function is integrable on S by the Fubini–Hobson–Tonelli theorem; the rest of the proof is immediate.   □

EXERCISES 5

A. Suppose that A and B are measurable subsets of $\mathbb{R}$ such that $m(A) = 0$. Show that the product $A \times B$ is a set of measure zero in $\mathbb{R}^2$.

B. Let φ be a step function on $S = [a, b] \times [c, d]$. Show that

$$\iint_S \varphi(x, y) \, dx \, dy = \int_a^b \left[\int_c^d \varphi(x, y) \, dy \right] dx = \int_c^d \left[\int_a^b \varphi(x, y) \, dx \right] dy.$$

C. Let f be integrable on $[0, 1] \times [0, 1]$. Show that

$$\int_0^1 \left[\int_0^x f(x, y) \, dy \right] dx = \int_0^1 \left[\int_y^1 f(x, y) \, dx \right] dy.$$

D. Let

$$f(x, y) = \begin{cases} \dfrac{x^2 - y^2}{(x^2 + y^2)^2} & \text{if } (x, y) \neq (0, 0), \\ 0 & \text{if } (x, y) = (0, 0). \end{cases}$$

(a) Show that

$$\int_0^1 \left[\int_0^1 f(x, y) \, dy \right] dx = \frac{\pi}{4},$$

$$\int_0^1 \left[\int_0^1 f(x, y) \, dx \right] dy = -\frac{\pi}{4}.$$

(b) Conclude from the Fubini theorem that f is not integrable on $[0, 1] \times [0, 1]$.

E. Let

$$f(x, y) = \begin{cases} \dfrac{x}{1 - y^2} & \text{if } |y| \neq 1, \\ 0 & \text{if } |y| = 1. \end{cases}$$

Show that

$$\int_{-1}^1 \left[\int_{-1}^1 f(x, y) \, dx \right] dy.$$

exists, while

$$\int_{-1}^1 \left[\int_{-1}^1 f(x, y) \, dy \right] dx$$

does not exist. *Hint*: If f is integrable on A, then f is integrable on every measurable subset of A.

F. Let $R = [-1, 1] \times [-1, 1]$ and let

$$f(x, y) = \begin{cases} \dfrac{xy}{(1 - |x|)^2 + (1 - |y|)^2} & \text{if } |xy| \neq 1, \\ 0 & \text{if } |xy| = 1. \end{cases}$$

(a) Show that

$$\int_{-1}^1 \left[\int_{-1}^1 f(x, y) \, dx \right] dy = \int_{-1}^1 \left[\int_{-1}^1 f(x, y) \, dy \right] dx = 0.$$

(b) Show that f is not integrable on R. *Hint*: f is not integrable on $[\frac{1}{2}, 1] \times [\frac{1}{2}, 1]$.

G. Let f and g be measurable functions on $[a, b]$. Show that the function $(x, y) \to f(x)g(y)$ is measurable on $[a, b] \times [a, b]$.

H. Compute the following integrals. *Hint*: See Exercise C:
   (a) $\int_0^1 [\int_y^1 \exp(-x^2) \, dx] \, dy$;
   (b) $\int_0^1 [\int_y^1 ((\sin x)/x) \, dx] \, dy$; and
   (c) $\int_0^1 [\int_y^1 \sin x^2 \, dx] \, dy$.

§6. The Complex Integral

So far we have restricted our attention to real-valued functions. In this section we will extend, in an obvious way, all our previous theory to complex-valued functions.

Recall first that any complex-valued function can be expressed in terms of two real functions by separating it into real and imaginary parts. In fact, if f is a complex-valued function, for each x in the domain of f, $f(x) \in \mathbb{C}$; thus, $f(x) = u_x + i v_x$, where u_x and v_x are real. The function $\operatorname{Re} f$ maps $x \to u_x$, while $\operatorname{Im} f$ maps $x \to v_x$. We write

$$f = \operatorname{Re} f + i \operatorname{Im} f.$$

The functions $\operatorname{Re} f$ and $\operatorname{Im} f$ are called, respectively, the *real part* and *imaginary part* of f. Since both $\operatorname{Re} f$ and $\operatorname{Im} f$ are real-valued functions, the following definition offers itself immediately from the definition of the integral for real-valued functions.

6.1. Definition. Let E be a measurable subset of $\mathbb{R}$. A function $f: E \to \mathbb{C}^*$ $(= \mathbb{C} \cup \{\infty\})$ is said to be *integrable* on E if both $\operatorname{Re} f$ and $\operatorname{Im} f$ are integrable. The integral of f is defined as

$$\int_E f(x)\, dx = \int_E \operatorname{Re} f(x)\, dx + i \int_E \operatorname{Im} f(x)\, dx.$$

It is easily shown that the space of all integrable complex-valued functions on E forms a complex vector space; that is, for any complex numbers α, β and for any complex-valued integrable functions f, g on E, the function $\alpha f + \beta g$ is again integrable on E.

We will now define the concept of a measurable function.

6.2. Definition. A function $f: E \to \mathbb{C}^*$ is said to be *measurable* on E if both $\operatorname{Re} f$ and $\operatorname{Im} f$ are measurable on E.

The inequality (see Proposition 4.5, Chapter II)

$$\left| \int_E f(x)\, dx \right| \le \int_E |f(x)|\, dx$$

for a real-valued integrable function f plays an important part in the theory of integration. The proof of the inequality for a complex-valued function requires more careful consideration. Once we have established the validity of the inequality for complex-valued functions, it follows easily that all those theorems of previous chapters continue to hold for complex-valued functions, excepting those whose statements become meaningless (e.g., the inequality $<$ between two complex numbers does not make sense). Recall that if $z = x + iy$ is a complex number, then $|z| = (x^2 + y^2)^{1/2}$.

6.3. Proposition. *Let $f: E \to \mathbf{C}^*$ be a measurable function. Then f is integrable on E if and only if $|f|$ is integrable on E; then*

$$\left| \int_E f(x) \, dx \right| \leq \int_E |f(x)| \, dx.$$

Proof. If f is integrable on E, then

$$|f| = [(\operatorname{Re} f)^2 + (\operatorname{Im} f)^2]^{1/2}$$

is measurable on E, and since

$$|f| \leq |\operatorname{Re} f| + |\operatorname{Im} f|,$$

$|f|$ is integrable by Proposition 1.5, Chapter III. Conversely, if $|f|$ is integrable, then since

$$|\operatorname{Re} f| \leq |f|, \qquad |\operatorname{Im} f| \leq |f|,$$

both $\operatorname{Re} f$ and $\operatorname{Im} f$ are integrable. Consequently, f is integrable on E.

Now we prove the inequality. Since f is integrable, we can write

$$\int_E f(x) \, dx = re^{i\theta}$$

for some real number $r \geq 0$ and $0 \leq \theta < 2\pi$. Let $c = e^{-i\theta}$. Then

$$c \int_E f(x) \, dx \geq 0.$$

Put $g = cf = u + iv$, where u and v are the real and imaginary parts of g, respectively. Then

$$\left| \int_E f(x) \, dx \right| = c \int_E f(x) \, dx = \int_E g(x) \, dx = \int_E u(x) \, dx$$

$$\leq \int_E |f(x)| \, dx. \qquad \qquad \square$$

We state the Beppo Levi theorem and the Lebesgue theorem for complex-valued functions without proofs.

6.4. The Beppo Levi Theorem. *Let (f_n) be a sequence of complex-valued integrable functions on E such that*

$$\sum_{n=1}^{\infty} \int_E |f_n(x)| \, dx < \infty.$$

Then $\sum_{n=1}^{\infty} f_n$ is integrable and

$$\int_E \left[\sum_{n=1}^{\infty} f_n(x) \right] dx = \sum_{n=1}^{\infty} \int_E f_n(x) \, dx.$$

6.5. The Lebesgue Theorem. *Let* (f_n) *be a sequence of complex-valued integrable functions on E such that* $\lim f_n(x)$ *exists almost everywhere on E. Suppose that there exists a real-valued integrable function g on E such that*

$$|f_n(x)| \le g(x)$$

for almost all x in E. Then the limit function $\lim f_n$ *is integrable on E and*

$$\int_E [\lim f_n(x)] \, dx = \lim \int_E f_n(x) \, dx.$$

EXERCISES 6

A. Prove the Beppo Levi theorem.

B. Prove the Lebesgue theorem.

C. The set of complex-valued integrable functions on a measurable set E of $\mathbb{R}$ is denoted by $L^1(E; \mathbb{C})$ (where equivalent functions are identified). Show that the L^1 norm makes $L^1(E; \mathbb{C})$ a Banach space.

D. *Fourier Transformations.* Recall that $e^{i\theta} = \cos\theta + i\sin\theta$ for $\theta \in \mathbb{R}$. Let $f \in L^1(\mathbb{R}; \mathbb{C})$.
   (a) Show that the integral

$$\hat{f}(x) = \int_{\mathbb{R}} e^{ixt} f(t) \, dt$$

   exists for all real x. ($\hat{f}$ is called the *Fourier transformation* of f.)
   (b) Show that $\sup\{|\hat{f}(x)|: x \in \mathbb{R}\} \le \|f\|$.

E. If (f_n) is a sequence in $L^1(\mathbb{R}; \mathbb{C})$ converging to f in the L^1 norm, then show that $(\hat{f}_n)$ converges uniformly to $\hat{f}$ on $\mathbb{R}$.

CHAPTER V
Differentiation and the Fundamental Theorem of Calculus

If f is integrable, we call the function

$$F(x) = \int_a^x f(t)\, dt \tag{1}$$

an *indefinite integral* of f. If any constant c is added to the right side of (1) the result

$$F(x) = \int_a^x f(t)\, dt + c$$

is also called an indefinite integral.

It is well known to the reader who has a rudimentary knowledge of calculus that differentiation and Riemann integration are inverse operations in the following sense:

(a) *If f is Riemann integrable on $[a, b]$, then its indefinite integral $F(x)$ is continuous on $[a, b]$; furthermore, if f is continuous at a point ξ in $[a, b]$, then F is differentiable at ξ and*

$$F'(\xi) = f(\xi). \tag{2}$$

(b) *If f is Riemann integrable on $[a, b]$ and if there is a differentiable function F on $[a, b]$ such that $F' = f$, then*

$$\int_a^x f(t)\, dt = F(x) - F(a). \tag{3}$$

This is the theorem which is usually called the *fundamental theorem of calculus*.

It is the purpose of this chapter to investigate the relationship of differentiation and integration as precisely as possible for Lebesgue integrable functions. In fact, we shall show that *relation* (2) *holds almost everywhere for a Lebesgue integrable function f*. This establishes that differentiation is the inverse operation of Lebesgue integration. However, relation (3) does not hold in general for Lebesgue integration. Three things might go wrong:

(i) the derivative $F'(x)$ might be undefined in a subset of $[a, b]$ which is not of measure zero;
(ii) $F'(x)$ might not be integrable on $[a, b]$; and
(iii) both sides of (3) might exist but differ.

Each of these possibilities can, in fact, be realized. We shall characterize a class of functions which satisfy relation (3).

§1. Nowhere Differentiable Functions

We recall that a function f, defined on $[a, b]$, is *differentiable at the point* x_0 in $[a, b]$ if the limit

$$\lim_{x \to x_0} \frac{f(x) - f(x_0)}{x - x_0}, \qquad a \le x \le b, \quad x \ne x_0, \tag{1}$$

exists. The limit is then called the *derivative* of f at x_0 and is denoted by $f'(x_0)$. In this way we obtain a function f' whose domain is a subset of $[a, b]$. If the domain of f' is the entire set $[a, b]$, we call f *differentiable on* $[a, b]$. Notice that no knowledge of the function outside the interval is needed.

It is possible to consider right-hand and left-hand limits in (1); this will lead to the definition of right-hand and left-hand derivatives. In particular, at the endpoints a and b, the derivative, if it exists, is a right- or left-hand derivative, respectively. We shall discuss a much more general concept of derivatives in the next section and shall not discuss here one-sided derivatives in any detail.

If f is defined on (a, b) and $x_0 \in (a, b)$, then $f'(x_0)$ is defined by (1) as above. In this case $f'(a)$ and $f'(b)$ are not defined. We notice that (a, b) can be unbounded.

A *function f having a derivative at the point* x_0 *is continuous at* x_0. This is a well-known fact. However, it is easy to see that the converse is not true. The example $f(x) = |x|$ shows that a function can be continuous without being differentiable. By taking simple algebraic combinations of the function $f(x) = |x|$ we can construct a continuous function which does not have a derivative at a finite or even a countably infinite number of points. However, the general opinion of mathematicians at the beginning of the nineteenth century was that a continuous function must have a substantial set of points at which it has a derivative.

The first serious consideration of this problem came in 1806, when A.M. Ampère tried to establish the differentiability of *any* function (continuous or not) except on a certain *negligible* set. Ampère's attempt was unsuccessful.

An end was put to these conjectures when Weierstrass (1815–1897) shocked the mathematical world by presenting before the Berlin Academy on July 18, 1872 an example of a function which is continuous at every point but whose derivative does not exist anywhere. His example was first published in Du Bois–Reymond (1875). His function is defined in the following way:

$$f(x) = \sum_{k=0}^{\infty} b^k \cos(a^k \pi x),$$

where a is an odd natural number and b a real number such that $0 < b < 1$ and $ab > 1 + 3\pi/2$. The Austrian mathematician and theologian Bernard Bolzano (1781–1848), who never taught mathematics in a professional capacity, had constructed a similar example in 1830. The manuscript was discovered only in 1920 and published in 1930 (a hundred years after it was written) in *Funktionenlehre*. Since Weierstrass, there have been many such examples constructed. Among them the following example seems simpler than any other the author has seen. It is a special case of one given in F.A. Behrend, "Crinkly curves and choppy surfaces" (1960).

We now describe a straightforward method of constructing a function which is continuous on ℝ but nowhere differentiable. The basic idea is to crinkle a curve infinitely. Let φ be a function such that

$$\varphi(x) = |x| \qquad \text{for } |x| \le 2,$$

$$\varphi(x + 4n) = \varphi(x), \qquad n = \pm 1, \pm 2, \ldots.$$

Then:

(1) φ is continuous.
(2) $0 \le \varphi(x) \le 2$; $|\varphi(x) - \varphi(y)| \le 2$.
(3) $-1 \le \dfrac{\varphi(x) - \varphi(y)}{x - y} \le 1$.
(4) For every c there is x_0 such that $|c - x_0| = 1$ and

$$\left| \frac{\varphi(c) - \varphi(x_0)}{c - x_0} \right| = 1.$$

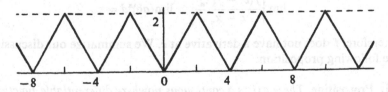

Figure 5.1

For $a > 0$ and $b > 0$ let

$$\varphi_n(x) = a^n \varphi(b^n x), \qquad n = 0, 1, 2, \ldots.$$

Then:

(5) φ_n is continuous.

(6) $0 \leq \varphi_n(x) \leq 2a^n$; $|\varphi_n(x) - \varphi_n(y)| \leq 2a^n$.

(7) $\left| \dfrac{\varphi_n(x) - \varphi_n(y)}{x - y} \right| \leq a^n b^n$.

(8) For every c there is x_n such that $|c - x_n| = 1/b^n$ and

$$\left| \frac{\varphi_n(c) - \varphi_n(x_n)}{c - x_n} \right| = a^n b^n.$$

We choose a to satisfy $0 < a < 1$ and set

$$f(x) = \sum_{n=0}^{\infty} \varphi_n(x).$$

Then we can show that the series converges uniformly because of (6) by the Weierstrass M-test. Hence f is continuous because it is the uniform limit of continuous functions.

To show that f is not differentiable, we assume $ab > 1$. Then for every c we have $x_n \to c$, and

$$\left| \frac{f(c) - f(x_n)}{c - x_n} \right| = \left| \sum_{j=0}^{\infty} \frac{\varphi_j(c) - \varphi_j(x_n)}{c - x_n} \right|$$

$$\geq \left| \frac{\varphi_n(c) - \varphi_n(x_n)}{c - x_n} \right| - \left(\sum_{j=0}^{n-1} + \sum_{j=n+1}^{\infty} \right) \left| \frac{\varphi_j(c) - \varphi_j(x_n)}{c - x_n} \right|$$

$$\geq (ab)^n - \sum_{j=0}^{n-1} (ab)^j - b^n \sum_{j=n+1}^{\infty} 2a^j \quad \text{[by (7) and (6)]}$$

$$= (ab)^n - \frac{(ab)^n - 1}{ab - 1} - \frac{2a(ab)^n}{1 - a}$$

$$> (ab)^n \left(1 - \frac{1}{ab - 1} - \frac{2a}{1 - a} \right).$$

If $d = 1 - 1/(ab - 1) - 2a/(1 - a) > 0$, then

$$\lim \left| \frac{f(c) - f(x_n)}{c - x_n} \right| > \lim (ab)^n d = \infty.$$

Therefore, f does not have a derivative at c. We summarize our discussion in the following proposition:

1.1. Proposition. *There exists a continuous nowhere differentiable function.*

It is also known that "most" continuous functions are nowhere differentiable. The argument is based on Baire's category theorem. In fact, the set of differentiable continuous functions in the Banach space of continuous functions on [0, 1] is of first category [see Boas, *A Primer of Real Functions* (1972), p. 61].

EXERCISES 1

Of the following, Exercises A, B, and C are for review.

A. **Rolle's Theorem.** *Suppose that f is continuous on $[a, b]$, that the derivative f' exists in the open interval (a, b), and that $f(a) = f(b) = 0$. Show that there exists a point c in (a, b) such that $f'(c) = 0$.*

B. **Mean Value Theorem.** *Suppose that f is continuous on $[a, b]$ and that f is differentiable on (a, b). Show that there exists a point c in (a, b) such that*

$$f(b) - f(a) = f'(c)(b - a).$$

C. **Taylor's Theorem.** *Let $n \in \mathbb{N}$. Suppose that f and its derivatives $f', f'', \ldots, f^{(n-1)}$ are defined and continuous on $[a, b]$ and that $f^{(n)}$ exists in (a, b). If x and y belong to $[a, b]$, then there exists a number z between x and y such that*

$$f(y) = f(x) + \frac{f'(x)}{1!}(y - x) + \frac{f''(x)}{2!}(y - x)^2 + \cdots$$

$$+ \frac{f^{(n-1)}(x)}{(n-1)!}(y - x)^{n-1} + \frac{f^{(n)}(z)}{n!}(y - x)^n.$$

D. *Van der Waerden's Example of a Continuous Nowhere Differentiable Function* (Van der Waerden, 1930). Let

$$\varphi_0(x) = \begin{cases} x & \text{if } 0 \leq x \leq \frac{1}{2}, \\ 1 - x & \text{if } \frac{1}{2} \leq x \leq 1. \end{cases}$$

Extend φ_0 by periodicity with period 1 to the whole line $\mathbb{R}$. Then let

$$\varphi_n(x) = \frac{\varphi_0(4^n x)}{4^n}, \qquad n = 1, 2, \ldots,$$

$$f(x) = \sum_{n=0}^{\infty} \varphi_n(x).$$

Show that:
(a) f is continuous everywhere; and
(b) f is nowhere differentiable.
Hint: Consider

$$\frac{f(x_0 \pm 4^{-n}) - f(x_0)}{4^{-n}}.$$

§2. The Dini Derivatives

We shall consider some generalizations of the usual derivative which have the advantage of applying to functions that are not necessarily differentiable (in the usual sense). These are the four Dini derivatives, which the Italian mathematician Ulisse Dini (1845–1918) introduced in *Fondamenti per la teorica della funzioni di variabili reali* (1878). Dini's derivatives may have either real values, or the values ∞ or $-\infty$. In order to define the Dini derivatives we shall write:

$$\sup E = \infty \quad \text{if the set E has no upper bound,}$$

$$\inf E = -\infty \quad \text{if the set E has no lower bound.}$$

Under this convention $\sup E$ and $\inf E$ always exist and have either real values or the values ∞ or $-\infty$.

2.1. Definition. Let $f: \mathbb{R} \to \mathbb{R}$. The *upper right, lower right, upper left,* and *lower left Dini derivatives* of f at x are, respectively, defined by

$$D^+ f(x) = \varlimsup_{h \downarrow 0} \frac{f(x + h) - f(x)}{h},$$

$$D_+ f(x) = \varliminf_{h \downarrow 0} \frac{f(x + h) - f(x)}{h},$$

$$D^- f(x) = \varlimsup_{h \uparrow 0} \frac{f(x + h) - f(x)}{h},$$

$$D_- f(x) = \varliminf_{h \uparrow 0} \frac{f(x + h) - f(x)}{h}.$$

The $+$ and $-$ refer to right and left, respectively, and their upper and lower positions refer to limit superior and limit inferior (see Definition 5.2, Chapter I).

Figure 5.2 illustrates a case in which all four Dini derivatives are finite and distinct at the given point x_0.

2.2. Proposition. *The Dini derivatives always exist (finite or infinite) for any function f, and*

$$D^+ f(x) \geq D_+ f(x), \qquad D^- f(x) \geq D_- f(x).$$

Proof. Let $x \in (a, b)$ and let

$$\ell(x) = \inf \left\{ \frac{f(x + h) - f(x)}{h} : 0 < h < b - x \right\},$$

$$\mathscr{L}(x) = \sup \left\{ \frac{f(x + h) - f(x)}{h} : 0 < h < b - x \right\}.$$

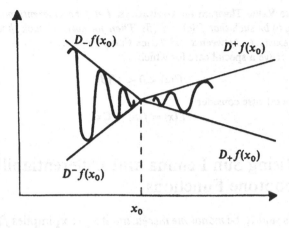

Figure 5.2

Then $\ell(x)$ and $\mathscr{L}(x)$ are functions of x and b. As b decreases to x, $\ell(x)$ increases and $\mathscr{L}(x)$ decreases. Consequently,

$$D^+ f(x) = \lim_{b \downarrow x} \mathscr{L}(x) \qquad \text{and} \qquad D_+ f(x) = \lim_{b \downarrow x} \ell(x)$$

exist (finite or infinite) and $D^+ f(x) \geq D_+ f(x)$. Similarly, we can show that both $D^- f(x)$ and $D_- f(x)$ exist (finite or infinite) and $D^- f(x) \geq D_- f(x)$. □

2.3. Proposition. *A function f is differentiable at x if and only if all four Dini derivatives are identical and are different from $\pm\infty$.*

Proof. The proof is left to the reader. □

EXERCISES 2

A. Prove Proposition 2.3.

B. Show that if $f'(x)$ exists (finite), then f is continuous at x.

C. Show that f may be discontinuous at x_0 when all four Dini derivatives at x_0 are identically equal to $+\infty$ or to $-\infty$.
   Hint:

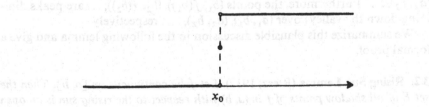

D. Intermediate Value Theorem for Derivatives. *Let* f *be differentiable on* (a, b) *and let* $\alpha, \beta \in (a, b)$ *be such that* $f'(\alpha) < f'(\beta)$. *Then for every* C *such that* $f'(\alpha) < C < f'(\beta)$ *there exists a* γ *between* α *and* β *such that* $f'(\gamma) = C$.
Hint: First prove a special case for which

$$f'(\alpha) < 0 < f'(\beta).$$

For the general case consider

$$F(x) = f(x) - Cx.$$

§3. The Rising Sun Lemma and Differentiability of Monotone Functions

A function f is said to be *monotone increasing* if $x_1 < x_2$ implies $f(x_1) \leq f(x_2)$ and *monotone decreasing* if $x_1 < x_2$ implies $f(x_1) \geq f(x_2)$. By a *monotone function* is meant a function which is either monotone increasing or monotone decreasing.

Some important properties of monotone functions are listed in Exercises 3A, 3B, and 3C.

Our primary aim in this section is to prove a celebrated theorem by Lebesgue (1904) which asserts that *every monotone function* f *defined on* $[a, b]$ *is differentiable almost everywhere*; that is,

$$-\infty < D_- f(x) = D^- f(x) = D_+ f(x) = D^+ f(x) < \infty$$

for almost all x in $[a, b]$. Lebesgue deduced this theorem from his entire theory of integration.

In this section we shall give an elementary proof due to F. Riesz which does not require any knowledge of Chapters II and III of this book. We follow the proof given in Riesz and Sz.-Nagy, *Functional Analysis* (1956).

3.1. Definition. Let f be a continuous function on $[a, b]$. A point x_0 in $[a, b]$ is called a *shadow point of* f *with respect to the rising sun* if there is a point ξ in $[a, b]$ such that $\xi > x_0$ and $f(\xi) > f(x_0)$.

The rationale for this terminology is indicated in Figure 5.3. The parallel lines are the rays of the sun rising in the east. We view the graph of f as a mountain ridge. It is intuitively clear that the set of all shadow points in (a, b) is an open set which is the union of pairwise disjoint open intervals (a_1, b_1), $(a_2, b_2), \ldots$. Furthermore, the points $(b_1, f(b_1)), (b_2, f(b_2)), \ldots$ are peaks dimming down the valleys over $(a_1, b_1), (a_2, b_2), \ldots$, respectively.

We summarize this plausible discussion in the following lemma and give a formal proof.

3.2. Rising Sun Lemma (Riesz, 1932). *Let* f *be continuous on* $[a, b]$. *Then the set* E *of all shadow points of* f *in* (a, b) *with respect to the rising sun is an open*

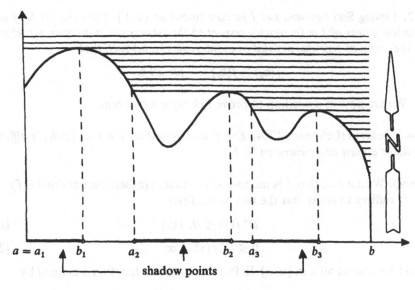

Figure 5.3

set which is the union of pairwise disjoint open intervals (a_k, b_k) such that

$$f(a_k) \le f(b_k) \qquad \text{for all } k.$$

Proof. We first demonstrate that E is open. Let $x_0 \in E$. Then there is a point $\xi > x_0$ with $f(\xi) > f(x_0)$. Since f is continuous, we can find a $\delta > 0$ such that if $x_0 - \delta < x < x_0 + \delta$, then $f(x) < f(\xi)$; that is, $(x_0 - \delta, x_0 + \delta) \subset E$. Thus E is open and therefore E is the union of a sequence of pairwise disjoint open intervals (a_k, b_k) (see Theorem 5.3, Chapter Zero).

We must show that $f(a_k) \le f(b_k)$. It suffices to prove that $f(x) \le f(b_k)$ for every $x \in (a_k, b_k)$, since by continuity of f at a_k we have $f(x) \to f(a_k)$ as $x \to a_k$; hence $f(a_k) \le f(b_k)$. For $x \in (a_k, b_k)$, let

$$A = \{y \in [x, b_k]: f(y) \ge f(x)\}.$$

Then A is a bounded nonempty set. Let $t = \sup A$. We claim that $t = b_k$. Suppose that $t < b_k$. Then $t \in (a_k, b_k)$; hence it is a shadow point. Therefore, there is a $\xi > t$ such that $f(\xi) > f(t)$. Since $t = \sup A$, this means that $\xi > b_k$ and $f(\xi) > f(b_k)$. Therefore, b_k is a shadow point, a contradiction. Thus $b_k = t$ and $f(x) \le f(b_k)$. $\square$

Analogously we can define *shadow points of f with respect to the setting sun*. Then virtually the same proof as that of the Rising Sun Lemma will show the following assertion:

3.3. Setting Sun Lemma. *Let f be continuous on* $[a, b]$. *Then the set E of all shadow points of f in* (a, b) *with respect to the setting sun is an open set which is the union of pairwise disjoint open intervals* (a_k, b_k) *such that*

$$f(a_k) \geq f(b_k) \quad \text{for all } k.$$

We are now in a position to prove Lebesgue's theorem.

3.4. Theorem (Lebesgue, 1904). *Every monotone function f on* $[a, b]$ *is differentiable almost everywhere on* $[a, b]$.

Proof. We assume that f is monotone increasing (otherwise consider $-f$).
It suffices to show that the two inequalities

$$D^+ f(x) \leq D_- f(x) \tag{1}$$

$$0 \leq D^+ f(x) < \infty \tag{2}$$

hold for almost all x in $[a, b]$. In fact, consider the function g defined by

$$g(x) = -f(-x).$$

Then g is monotone increasing on $[-b, -a]$. Furthermore, it can be shown easily that

$$D_- g(-x) = D_+ f(x), \qquad D^+ g(-x) = D^- f(x). \tag{3}$$

(See Exercise 3E). Hence, applying (1) to g and combining it with (3), we get

$$D^- f(x) \leq D_+ f(x) \tag{4}$$

almost everywhere. Since

$$D^+ f(x) \geq D_+ f(x), \qquad D^- f(x) \geq D_- f(x)$$

are always true for any x, we have from (1), (2), and (4)

$$0 \leq D^+ f(x) \leq D_- f(x) \leq D^- f(x) \leq D_+ f(x) \leq D^+ f(x) < \infty$$

for almost all x in $[a, b]$; hence the equality signs must hold, which was to be proved. Therefore, it remains to prove the inequalities (1) and (2). We shall prove these inequalities in the following lemmas. $\qquad \square$

In the following sequence of lemmas we assume that f is a continuous and monotone increasing function on $[a, b]$:

3.5. Lemma. *For any real number* $r > 0$, *and for any open interval* $(\alpha, \beta) \subset (a, b)$, *the set*

$$E^r = \{x \in (\alpha, \beta): D^+ f(x) > r\}$$

can be covered by a sequence of pairwise disjoint open intervals of total length less than or equal to

$$\frac{1}{r}[f(\beta) - f(\alpha)].$$

Proof. Let $x_0 \in (\alpha, \beta)$ be such that $D^+ f(x_0) > r$. Then there exists a point $\xi > x_0$ such that

$$\frac{f(\xi) - f(x_0)}{\xi - x_0} > r,$$

or

$$f(\xi) - r\xi > f(x_0) - rx_0.$$

Therefore, x_0 is a shadow point of the function $f(x) - rx$ with respect to the rising sun. Hence, by the Rising Sun Lemma, E^r is contained in the union of a sequence of pairwise disjoint open intervals $(a_k, b_k) \subset (\alpha, \beta)$ for which

$$f(a_k) - ra_k \le f(b_k) - rb_k,$$

or

$$b_k - a_k \le \frac{1}{r}[f(b_k) - f(a_k)].$$

The total length of these intervals is equal to

$$\sum_{k=1}^{\infty} (b_k - a_k) \le \frac{1}{r} \sum_{k=1}^{\infty} [f(b_k) - f(a_k)] \le \frac{1}{r}[f(\beta) - f(\alpha)]. \qquad \square$$

3.6. Lemma. *For any real number $r > 0$, and for any open interval $(\alpha, \beta) \subset (a, b)$, the set*

$$E_r = \{x \in (\alpha, \beta): D_- f(x) < r\}$$

can be covered by a sequence of pairwise disjoint open intervals $(\alpha_k, \beta_k) \subset (\alpha, \beta)$ such that

$$f(\beta_k) - f(\alpha_k) \le r(\beta_k - \alpha_k).$$

Proof. Let $x_0 \in E_r$. Then $D_- f(x_0) < r$, and hence there is a point $\xi < x_0$ such that

$$\frac{f(\xi) - f(x_0)}{\xi - x_0} < r,$$

or

$$f(\xi) - r\xi > f(x_0) - rx_0.$$

This means that x_0 is a shadow point of $f(x) - rx$ with respect to the setting sun. Therefore, by the Setting Sun Lemma, E_r is covered by countably many pairwise disjoint open intervals $(\alpha_k, \beta_k) \subset (\alpha, \beta)$ such that

$$f(\alpha_k) - r\alpha_k \ge f(\beta_k) - r\beta_k,$$

or

$$f(\beta_k) - f(\alpha_k) \le r(\beta_k - \alpha_k). \qquad \square$$

Combining Lemmas 3.5 and 3.6 we have the following:

3.7. Lemma. *Let $0 < r < R < \infty$. Then for any open interval $(\alpha, \beta) \subset (a, b)$, the set*

$$E_r^R = \{x \in (\alpha, \beta): D_- f(x) < r < R < D^+ f(x)\}$$

can be covered by a sequence of pairwise disjoint open intervals of total length less than or equal to

$$\frac{r}{R}(\beta - \alpha).$$

Proof. Notice that $E_r^R = E_r \cap E^R$. Therefore, by Lemma 3.6, E_r^R is covered by a sequence of pairwise open intervals (α_k, β_k) satisfying

$$f(\beta_k) - f(\alpha_k) \le r(\beta_k - \alpha_k). \qquad (*)$$

For each k we now consider $E_r^R \cap (\alpha_k, \beta_k)$. Then

$$E_r^R \cap (\alpha_k, \beta_k) \subset \{x \in (\alpha_k, \beta_k): D^+ f(x) > R\}.$$

Therefore, by Lemma 3.5, $E_r^R \cap (\alpha_k, \beta_k)$ can be covered by a sequence of pairwise disjoint open intervals of total length less than or equal to $(1/R)[f(\beta_k) - f(\alpha_k)]$. Therefore, the set E_r^R can be covered by a sequence of pairwise disjoint open intervals of total length less than or equal to

$$\frac{1}{R} \sum_{k=1}^{\infty} [f(\beta_k) - f(\alpha_k)].$$

But, by $(*)$, we have

$$\frac{1}{R} \sum_{k=1}^{\infty} [f(\beta_k) - f(\alpha_k)] \le \frac{r}{R} \sum_{k=1}^{\infty} (\beta_k - \alpha_k) \le \frac{r}{R}(\beta - \alpha). \qquad \square$$

We are now in a position to verify the inequalities (1) and (2) in the proof of Theorem 3.4.

3.8. Lemma. *The inequalities:*

(1) $D^+ f(x) \le D_- f(x)$; *and*
(2) $0 \le D^+ f(x) < \infty$;

hold for almost all x in $[a, b]$.

Proof. To prove inequality (1), it is sufficient to show that the set

$$E_r^R = \{x \in (a, b): D_- f(x) < r < R < D^+ f(x)\}$$

has measure zero, since a countable union of sets of measure zero is also of measure zero, and since

$$\{x \in (a, b): D_- f(x) < D^+ f(x)\} = \bigcup E_r^R,$$

where the union is taken over all rationals $R > r > 0$.

By Lemma 3.7, E_r^R can be covered by a sequence of pairwise disjoint open intervals (a_k, b_k) such that

$$\sum_{k=1}^{\infty} (b_k - a_k) \le \frac{r}{R}(b - a). \qquad (*)$$

Applying the same lemma again to $E_r^R \cap (a_k, b_k)$, we can assert that $E_r^R \cap (a_k, b_k)$ can be covered by a sequence of pairwise disjoint open intervals (a_{kn}, b_{kn}) such that

$$\sum_{n=1}^{\infty} (b_{kn} - a_{kn}) \leq \frac{r}{R}(b_k - a_k). \qquad (**)$$

It follows from $(*)$ and $(**)$ that

$$\sum_{k=1}^{\infty} \left[\sum_{n=1}^{\infty} (b_{kn} - a_{kn}) \right] \leq \left(\frac{r}{R} \right)^2 (b - a),$$

that is, the set E_r^R is covered by a sequence of pairwise disjoint open intervals of total length less than or equal to $(r/R)^2(b - a)$. Inductively, we conclude that for any natural number n, E_r^R can be covered by a sequence of pairwise disjoint open intervals of total length less than or equal to $(r/R)^n(b - a)$. Since $0 < r/R < 1$, $(r/R)^n \to 0$ as $n \to \infty$; hence E_r^R is of measure zero.

To prove inequality (2), we must show that the set

$$E = \{x \in (a, b): D^+ f(x) = \infty\}$$

has measure zero. Note that

$$E = \bigcap_{n=1}^{\infty} E^n,$$

where

$$E^n = \{x \in (a, b): D^+ f(x) > n\}.$$

We now apply Lemma 3.5 to each E^n and conclude that for each n, E is covered by a sequence of pairwise disjoint open intervals of total length less than or equal to

$$\frac{1}{n}[f(b) - f(a)].$$

It follows that E has measure zero. It remains to show that $D^+ f(x) \geq 0$; but this is obvious since f is monotone increasing.   □

Thus the theorem is proved in the case where the monotone function f is continuous. To extend this to the case of discontinuous functions, we need only a generalized Rising Sun Lemma, as indicated in Exercise 3F, noting that the remainder of the proof continues to carry through.

However, the proof can be troublesome and tedious because of the complexity of the analytic definition of a shadow point for a discontinuous function as described in Exercise 3F. Fortunately we can avoid this clumsiness. We present the following adaptation of the elegant proof that Theorem 3.4 holds for a discontinuous monotone function given by Lee A. Rubel in "Differentiability of monotone functions" (1963). The proof consists of the following three lemmas:

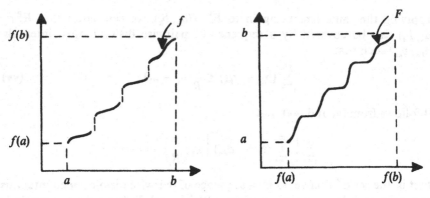

Figure 5.4

3.9. Lemma. *Let f be a strictly increasing function on $[a, b]$. Then f has a continuous inverse; that is, there exists a continuous, monotone increasing function F on $[f(a), f(b)]$ such that*

$$F(f(x)) = x \qquad \text{for each} \quad x \in [a, b].$$

Proof. Geometrically, the construction of F is evident (see Figure 5.4). Analytically, we define F by

$$F(y) = \sup\{t: f(t) \leq y\}.$$

It is clear that $F(f(x)) = x$ for all x in $[a, b]$ and F is monotone increasing on $[f(a), f(b)]$. Furthermore, F is continuous on $[f(a), f(b)]$, since $F([f(a), f(b)]) = [a, b]$. $\qquad\square$

It is convenient to denote $f'(x) = \infty$ if all the Dini derivatives of f at x are equal to ∞.

3.10. Lemma. *Let f be a monotone increasing function on $[a, b]$. Then $f'(x) \leq \infty$ for almost all x in $[a, b]$.*

Proof. We may assume that f is strictly increasing and satisfies

$$f(y) - f(x) \geq y - x$$

whenever $y \geq x$, since we could otherwise consider $f(x) + x$. By Lemma 3.9, let F be the continuous inverse of f defined on $[f(a), f(b)]$. Then $F'(y) < \infty$ holds for almost all y in $[f(a), f(b)]$. We write

$$\frac{f(y) - f(x)}{y - x} = \frac{f(y) - f(x)}{F(f(y)) - F(f(x))} = \left[\frac{F(f(y)) - F(f(x))}{f(y) - f(x)}\right]^{-1}.$$

Since every monotone function has at most countably many discontinuities,

$f(y) \to f(x)$ as $y \to x$ for almost all x in $[a, b]$. Therefore,

$$f'(x) = \lim_{y \to x} \frac{f(y) - f(x)}{y - x} \leq \infty \qquad \text{for almost all} \quad x \in [a, b]. \qquad \Box$$

We now show that every monotone function on $[a, b]$ is differentiable almost everywhere on $[a, b]$. This follows from the following lemma:

3.11. Lemma. *Let f be a monotone increasing function on $[a, b]$. Then*

$$E_\infty = \{x \in [a, b]: f'(x) = \infty\}$$

is of measure zero.

Proof. We assume that f is strictly increasing and satisfies

$$f(y) - f(x) \geq y - x$$

whenever $y \geq x$. Let $x \in E_\infty$. Then $D^+ f(x) = D^- f(x) = \infty$. Therefore, for every $C > 0$ there exist s and t with $s < x < t$ such that

$$f(t) - f(x) > C(t - x),$$

$$f(x) - f(s) > C(x - s),$$

[because $D^+ f(x) > C, D^- f(x) > C$]. Therefore

$$f(t) - f(s) > C(t - s).$$

Let E_C be the set of points x in (a, b) for which there exist s_x and t_x in (a, b) with $s_x < x < t_x$ such that

$$f(t_x) - f(s_x) > C(t_x - s_x). \tag{1}$$

Then $E_\infty \subset E_C$ for all $C > 0$. It is clear that E_C is open, and hence E_C is the union of pairwise disjoint intervals (a_n, b_n); i.e.,

$$E_C = \bigcup_{n=1}^{\infty} (a_n, b_n).$$

For each n, let $(a'_n, b'_n) \subset (a_n, b_n)$ be such that

$$2(b'_n - a'_n) = b_n - a_n. \tag{2}$$

Then the compact interval $[a'_n, b'_n]$ will be covered by the open intervals (s_x, t_x) for $x \in [a'_n, b'_n]$ and $(s_x, t_x) \subset (a_n, b_n)$; hence there is a finite sub-covering, say (s_k, t_k), where $k = 1, 2, \ldots, N$. We can assume that each point of $\bigcup_{k=1}^{N} (s_k, t_k)$ lies in at most two of the intervals, because given any three open intervals with a common point, some one must be contained in the union of the other two. Now we can break the family of the open intervals (s_k, t_k) into two subfamilies of disjoint intervals. Therefore we obtain

$$\sum_{k=1}^{N} [f(t_k) - f(s_k)] \leq 2[f(b_n) - f(a_n)]. \tag{3}$$

Hence,

$$b_n' - a_n' \leq \sum_{k=1}^{N} (t_k - s_k)$$

$$< \frac{1}{C} \sum_{k=1}^{N} [f(t_k) - f(s_k)] \quad \text{by (1)}$$

$$\leq \frac{2}{C} [f(b_n) - f(a_n)] \qquad \text{by (3)}.$$

Therefore,

$$\sum_{n=1}^{\infty} (b_n - a_n) = \sum_{n=1}^{\infty} 2(b_n' - a_n')$$

$$< \frac{4}{C} \sum_{n=1}^{\infty} [f(b_n) - f(a_n)]$$

$$\leq \frac{4}{C} [f(b) - f(a)].$$

Since $E_\infty \subset E_C$ and C is arbitrary, it follows that E_∞ is of measure zero. $\quad \square$

The theorem may be proved in a quite different way using what is known as the Vitali covering theorem. See, for instance, Natanson (1955), pp. 208–212. Lebesgue established the theorem for continuous monotone functions. Subsequently, G. Faber in 1910, G.C. Young and W.H. Young in 1911, and F. Riesz in 1932 gave proofs without the assumption of continuity.

EXERCISES 3

A. Let f be monotone increasing on $[a, b]$. Then $f(x^+)$ and $f(x^-)$ exist at every point x of (a, b). More precisely, we have the following relation:

$$\sup_{a < t < x} f(t) = f(x^-) \leq f(x) \leq f(x^+) = \inf_{x < t < b} f(t).$$

It is also true that if $a < x < y < b$, then

$$f(x^+) \leq f(y^-).$$

B. Let f be a monotone function on $[a, b]$. Then the set of points of $[a, b]$ at which f is discontinuous is at most countable.

C. Given a sequence (x_n) in $[a, b]$, construct a function $f: [a, b] \to \mathbb{R}$ which is monotone on $[a, b]$ and discontinuous at every point x_n and at no other point of $[a, b]$. *Hint:* Let (c_n) be a sequence of positive numbers such that $\sum_{n=1}^{\infty} c_n < \infty$. Define

$$f(x) = \sum_{x_n < x} c_n.$$

(The summation is to indicate that we sum over those indices n for which $x_n < x$.) Verify that:
(a) f is monotone increasing on $[a, b]$;
(b) f is discontinuous at each x_n by noticing $f(x_n^+) - f(x_n^-) = c_n$; and
(c) f is continuous at every other point of $[a, b]$.

D. Verify that in the statement of the Rising Sun Lemma we have exactly

$$f(a_k) = f(b_k)$$

except possibly when $a_k = a$.

E. Let $g(x) = -f(-x)$. Show that

$$D_- g(-x) = D_+ f(x), \qquad D^+ g(-x) = D^- f(x).$$

F. Prove Theorem 3.4 for a discontinuous function. To do this, modify the Rising Sun Lemma as indicated below.
   (1) f is continuous, except for jumps.
   (2) A point x_0 is a *shadow point* of f with respect to the rising sun if there is a point $\xi > x_0$ such that

   $$\max\{f(x_0^-), f(x_0), f(x_0^+)\} < f(\xi).$$

   (3) The inequality in the lemma is replaced by

   $$f(a_k^+) \le \max\{f(b_k^-), f(b_k), f(b_k^+)\}.$$

G. Let f be continuous on $[a, b]$ such that $D^+ f(x) > 0$ holds on (a, b). Show that $f(a) \le f(b)$.

H. Suppose that $E \subset (a, b)$ has measure zero. Is it possible to find a monotone function on $[a, b]$ which is differentiable on the complement of E?

§4. Functions of Bounded Variation

In this section we shall extend the Lebesgue theorem on differentiability of monotone functions to a larger class of functions, namely the class of functions of bounded variation. These functions are essentially differences of monotone increasing functions. Functions of bounded variation are important not only in differentiation but also in the study of Fourier series (see Jordan's Test, §9, Chapter VI), rectifiable curves, Riemann–Stieltjes integration, and functional analysis. We are indebted to Camille Jordan (1838–1922) for the definition and a complete characterization of functions of bounded variation [see C. Jordan, "Sur la série de Fourier" (1881)]. Jordan introduced this concept in his study of the Dirichlet theorem concerning convergence of Fourier series (see Theorem 2.2, Chapter I; Jordan's Test, §9, Chapter VI).

4.1. Definition. A function $f: [a, b] \to \mathbf{R}$ is said to be of *bounded variation* if there is a constant $C > 0$ such that the inequality

$$\sum_{k=1}^{n} |f(x_k) - f(x_{k-1})| \le C$$

holds for any partition

$$a = x_0 < x_1 < \cdots < x_n = b.$$

In this case, we define the *total variation* $V_a^b(f)$ of f on $[a, b]$ by

$$V_a^b(f) = \sup\left\{\sum_{k=1}^{n} |f(x_k) - f(x_{k-1})|: a = x_0 < x_1 < \cdots < x_n = b\right\},$$

where the supremum is taken over all partitions of $[a, b]$.

If f is a monotone function on $[a, b]$, then f is of bounded variation, and

$$V_a^b(f) = |f(b) - f(a)|.$$

It is easy to show that the sum and difference of two monotone increasing functions are of bounded variation. In fact, we have the following assertion.

4.2. Proposition. *Let f and g be functions of bounded variation on $[a, b]$. Then $f + g$ and fg are of bounded variation on $[a, b]$.*

Proof. For any partition

$$a = x_0 < x_1 < \cdots < x_n = b,$$

we have

$$\sum_{k=1}^{n} |f(x_k) - f(x_{k-1}) + g(x_k) - g(x_{k-1})|$$

$$\leq \sum_{k=1}^{n} |f(x_k) - f(x_{k-1})| + \sum_{k=1}^{n} |g(x_k) - g(x_{k-1})|$$

$$\leq V_a^b(f) + V_a^b(g).$$

Hence the sum $f + g$ is of bounded variation.

Next, let A and B be the suprema of f and g on $[a, b]$, respectively. These exist since every function of bounded variation is bounded (see Exercise 4B). If $h = fg$, then

$$h(x_k) - h(x_{k-1}) = f(x_k)[g(x_k) - g(x_{k-1})] + g(x_{k-1})[f(x_k) - f(x_{k-1})].$$

Hence

$$\sum_{k=1}^{n} |h(x_k) - h(x_{k-1})| \leq AV_a^b(g) + BV_a^b(f)$$

which proves that fg is of bounded variation. $\square$

4.3. Corollary. *If f and g are monotone increasing on $[a, b]$, then $f - g$ is of bounded variation on $[a, b]$.*

The converse of this corollary is also true. To show this fact, we need the following lemma:

4.4. Lemma. *Let f be of bounded variation on $[a, b]$.*

(a) *If $a \leq c \leq b$, then f is of bounded variation on each of $[a, c]$ and $[c, b]$. Moreover,*
$$V_a^b(f) = V_a^c(f) + V_c^b(f).$$

(b) *The function $x \to V_a^x(f)$ is monotone increasing on $[a, b]$.*

Proof. (a) It is clear that f is of bounded variation on $[a, c]$ and $[c, b]$. We may assume that $a < c < b$, since $V_a^a(f) = 0$. Let $\varepsilon > 0$ be given. Then there is a partition
$$a = x_0 < x_1 < \cdots < x_n = b$$
such that
$$V_a^b(f) - \varepsilon < \sum_{k=1}^{n} |f(x_k) - f(x_{k-1})| \leq V_a^b(f). \tag{$*$}$$

If c is not one of the x_k's, we add c to $\{x_0, x_1, \ldots, x_n\}$ and get a new partition for which ($*$) still holds (why?). We can therefore assert that
$$V_a^b(f) - \varepsilon \leq V_a^c(f) + V_c^b(f) \leq V_a^b(f).$$

Since ε was arbitrary, we have
$$V_a^b(f) = V_a^c(f) + V_c^b(f).$$

(b) The function $x \to V_a^x(f)$ is obviously monotone increasing. $\square$

We are now ready to characterize a function of bounded variation.

4.5. Jordan Decomposition Theorem. *Every function of bounded variation on $[a, b]$ is the difference of two monotone increasing functions on $[a, b]$.*

Proof. Write
$$f(x) = V_a^x(f) - [V_a^x(f) - f(x)].$$

Since the function $x \to V_a^x(f)$ is obviously monotone increasing, it remains to show that the function
$$x \to V_a^x(f) - f(x)$$
is monotone increasing. But if $x < y$, then
$$[V_a^y(f) - f(y)] - [V_a^x(f) - f(x)] = V_x^y(f) - [f(y) - f(x)] \geq 0,$$
since $|f(y) - f(x)| \leq V_x^y(f)$. $\square$

The decomposition of a function of bounded variation as the difference of two monotone increasing functions is, of course, not unique. For if f is represented by the difference of two monotone increasing functions g and h, and if p is a monotone increasing function, then f is also represented by the difference of monotone increasing functions $g + p$ and $h + p$ since

$$f = g - h = (g + p) - (h + p).$$

We can now state the substance of the Lebesgue theorem for differentiation in the following general form as a corollary of the Jordan decomposition theorem:

4.6. Theorem (Lebesgue, 1904). *Every function of bounded variation is differentiable almost everywhere.*

If f is of bounded variation on $[a, b]$, then f is differentiable almost everywhere on $[a, b]$. It is natural to ask if the derivative f' is integrable on $[a, b]$.

4.7. Proposition. *If f is a monotone increasing function on $[a, b]$, then the derivative f' is integrable and*

$$\int_a^b f'(x)\, dx \le f(b) - f(a).$$

Proof. We extend the definition of f to the interval $[a, b + 1]$ by the relation $f(x) = f(b)$ if $b < x \le b + 1$. Now let

$$f_n(x) = n\left[f\left(x + \frac{1}{n}\right) - f(x)\right]$$

for each $n \in \mathbb{N}$ and $x \in [a, b]$. It is clear that $f_n(x) \to f'(x)$ as $n \to \infty$ for almost all $x \in [a, b]$; hence f' is measurable. Since f is integrable on $[a, b]$, so is each f_n. Furthermore, $f_n \ge 0$. Integrating f_n, we get

$$\int_a^b f_n(x)\, dx = n\int_a^b\left[f\left(x + \frac{1}{n}\right) - f(x)\right] dx$$

$$= n\left[\int_{a+(1/n)}^{b+(1/n)} f(x)\, dx - \int_a^b f(x)\, dx\right]$$

$$= n\left[\int_b^{b+(1/n)} f(x)\, dx - \int_a^{a+(1/n)} f(x)\, dx\right]$$

$$\le f(b) - f(a).$$

Therefore, by Theorem 6.6 (Fatou's lemma), Chapter II, we conclude that f' is integrable and

$$\int_a^b f'(x)\, dx \le f(b) - f(a). \qquad \square$$

4.8. Corollary. *If f is of bounded variation on $[a, b]$, then the derivative f' is integrable on $[a, b]$.*

We notice that the equality in Proposition 4.7 need not hold in general; that is, there is a monotone increasing function f satisfying

$$\int_a^b f'(x)\, dx < f(b) - f(a).$$

For instance, let f be a step function. Then its derivative vanishes almost everywhere. It turns out that the strict inequality in Proposition 4.7 can arise in practice even for continuous monotone increasing functions. We shall exhibit a continuous monotone increasing function $f: [0, 1] \to \mathbb{R}$ such that $f(0) = 0$, $f(1) = 1$, and $f'(x) = 0$ almost everywhere.

4.9. Example. In Exercise 4G, Chapter I, we defined the Lebesgue singular function analytically. (The term "singular" is used for a continuous function whose derivative vanishes almost everywhere.) In Figure 5.5 we give a brief graphic description of this function.

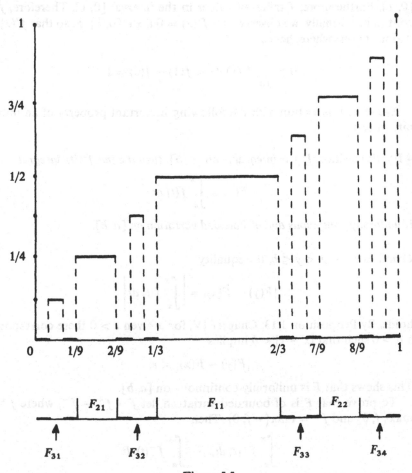

Figure 5.5

Let F be the Cantor ternary set constructed in Example 4.4, Chapter I. We use the notations established in this example. For each n, let

$$F_{n1}, F_{n2}, \ldots, F_{n2^{n-1}}$$

be the open intervals removed in the process to obtain F_n from F_{n-1}. We define a function $f: [0, 1]\backslash F \to \mathbb{R}$ by

$$f(x) = \frac{2k - 1}{2^n}$$

if $x \in F_{nk}$ for some n and $k = 1, 2, \ldots, 2^{n-1}$. Then it is easy to see that f is monotone increasing on $[0, 1]\backslash F$. We now extend f to the entire interval $[0, 1]$. First we define $f(0) = 0$, $f(1) = 1$. If $x \in F, 0 < x < 1$, then there is an increasing sequence (x_n) in $[0, 1]\backslash F$ such that $x_n \to x$ (why?). The limit of $(f(x_n))$, say $f(x)$, exists, and it is easy to show that $f(x)$ is independent of the choice of (x_n). It is then obvious that f is a monotone increasing function on $[0, 1]$. Furthermore, f takes all values in the interval $[0, 1]$. Therefore, f is continuous. Finally, we observe that $f'(x) = 0$ if $x \in [0, 1]\backslash F$, so that $f'(x) = 0$ almost everywhere; hence

$$0 = \int_0^1 f'(x)\, dx < f(1) - f(0) = 1. \qquad \square$$

We close this section with the following important property of an indefinite integral:

4.10. Proposition. *If f is integrable on $[a, b]$, then the indefinite integral*

$$F(x) = \int_a^x f(t)\, dt$$

is uniformly continuous and of bounded variation on $[a, b]$.

Proof. For $a \le x < y \le b$, the equality

$$|F(y) - F(x)| = \left| \int_x^y f(t)\, dt \right|$$

holds. By Proposition 1.13, Chapter IV, for a given $\varepsilon > 0$ there corresponds a $\delta > 0$ such that $|x - y| < \delta$ implies

$$|F(y) - F(x)| < \varepsilon.$$

This shows that F is uniformly continuous on $[a, b]$.

To prove that F is of bounded variation, let $f = f^+ - f^-$, where $f^+ = \max\{f, 0\}$ and $f^- = \max\{-f, 0\}$. Then

$$\int_a^x f^+(t)\, dt, \qquad \int_a^x f^-(t)\, dt$$

are monotone increasing functions of x, and we have

$$F(x) = \int_a^x f^+(t)\, dt - \int_a^x f^-(t)\, dt.$$

Therefore, F is of bounded variation by the Jordan decomposition theorem. □

By what we have proved, the indefinite integral $F(x) = \int_a^x f(t)\, dt$ is differentiable almost everywhere. We shall establish in §6 that $F'(x) = f(x)$ for almost all $x \in [a, b]$.

EXERCISES 4

A. Show that the Dirichlet function $D: [0, 1] \to \{0, 1\}$, which takes the value 1 on rationals and 0 on irrationals, is not of bounded variation.

B. Show that if f is of bounded variation on $[a, b]$, then f is bounded on $[a, b]$.

C. Show that a continuous function may not be of bounded variation.

D. Suppose that f' exists everywhere and is bounded on $[a, b]$. Show that f is of bounded variation. *Hint:* The mean value theorem.

E. Show that the continuous function

$$f(x) = \begin{cases} x^\alpha \sin \pi/x^\beta & \text{if } x \neq 0, \\ 0 & \text{if } x = 0, \end{cases}$$

is of bounded variation on $[0, 1]$ if $\alpha > \beta$. *Hint:* Exercise D.

F. Show that the function f in Exercise E is not of bounded variation if $\alpha \leq \beta$.

G. Show that the set of discontinuities of a function of bounded variation is at most countable.

H. *Rectifiable Curves.* A curve $y = f(x)$ on $[a, b]$ is said to be *rectifiable* if the length of the polygonal lines with successive vertices at the points $(x_0, f(x_0)), \ldots,$ $(x_n, f(x_n))$ is bounded by a constant for every partition $a = x_0 < x_1 < \cdots < x_n = b$. Show that a curve $y = f(x)$ is rectifiable if and only if the function f is of bounded variation.

I. Denote by $BV_0[a, b]$ the space of all functions f of bounded variation on $[a, b]$ such that $f(a^+) = 0$ [$f(a^+)$ exists for every f of bounded variation by Exercise 3A]. For $f \in BV_0[a, b]$, define $\|f\|$ by

$$\|f\| = V_a^b(f).$$

Show that $\|f\|$ is a norm on $BV_0[a, b]$.

J. Show directly that if f is Riemann integrable on $[a, b]$, then its indefinite integral is uniformly continuous on $[a, b]$.

§5. Absolute Continuity

In this section we introduce a new class of functions, which plays a fundamental role in clarifying the relation between differentiation and integration
in the Lebesgue sense. In fact, we want to characterize a class of functions f
satisfying the fundamental relation

$$\int_a^x f'(x)\, dx = f(x) - f(a). \tag{*}$$

It is clear that such a function should be uniformly continuous and be of
bounded variation by virtue of Proposition 4.10. But it is known that the
relation (*) is not necessarily true even when f is monotone increasing and
is continuous on $[a, b]$ (see Example 4.9). Thus we must impose stronger
conditions than continuity or bounded variation to ensure the relation (*).

5.1. Definition. A function $f: [a, b] \to \mathbb{R}$ is said to be *absolutely continuous*
if for any $\varepsilon > 0$ there exists a $\delta > 0$ such that for every finite collection of pairwise disjoint intervals $(a_k, b_k) \subset [a, b]$, $k = 1, 2, \ldots, n$, with
$\sum_{k=1}^n (b_k - a_k) < \delta$, we have

$$\sum_{k=1}^n |f(b_k) - f(a_k)| < \varepsilon.$$

Cleary, *every absolutely continuous function f is uniformly continuous.* In
fact, $|f(b_1) - f(a_1)|$ is small for every sufficiently small single interval (a_1, b_1)
in $[a, b]$. Thus, absolute continuity is at least as strong as uniform continuity.
It is, moreover, stronger, since there are uniformly continuous functions
which are not absolutely continuous (see Exercise 5B). The concept of absolute continuity is due to Vitali (1908).

It is easy to show that the absolute continuity of a function can be defined
without requiring the finiteness of the collection of pairwise disjoint intervals;
that is, a function is *absolutely continuous* on $[a, b]$ if and only if for an
arbitrary $\varepsilon > 0$ there exists a $\delta > 0$ such that for any sequence of pairwise
disjoint intervals $(a_k, b_k) \subset [a, b]$ of total length less than δ, we have

$$\sum_{k=1}^\infty |f(b_k) - f(a_k)| < \varepsilon.$$

There are various simple sufficient conditions for a function to be absolutely continuous.

A function $f: [a, b] \to \mathbb{R}$ is said to satisfy a *Lipschitz condition* on $[a, b]$ [in
honor of the German mathematician Rudolf Lipschitz (1831–1904)] if there
is a constant $M > 0$ such that

$$|f(x) - f(y)| \le M|x - y|$$

for all x and y in $[a, b]$.

5.2. Proposition.

(a) *A function satisfying a Lipschitz condition is absolutely continuous.*
(b) *A function f satisfies a Lipschitz condition on [a, b] if and only if f is differentiable on [a, b] and the derivative f' is bounded.*

Proof. The first part is clear. For the second part, use the mean value theorem for derivatives.   □

5.3. Proposition. *If f is absolutely continuous on [a, b], then so is αf, where α is a constant. Moreover, if f and g are absolutely continuous on [a, b], then so are f + g and fg.*

Proof. Exercise 5E.   □

5.4. Proposition. *If f is absolutely continuous on [a, b], then f is of bounded variation on [a, b].*

Proof. For $\varepsilon = 1$ there exists a $\delta > 0$ such that for every finite collection of pairwise disjoint intervals $(a_k, b_k) \subset [a, b]$, $k = 1, 2, \ldots, n$, with

$$\sum_{k=1}^{n} (b_k - a_k) < \delta$$

we have

$$\sum_{k=1}^{n} |f(b_k) - f(a_k)| < 1.$$

Hence if $[c, d]$ is any subinterval of $[a, b]$ of length less than δ, we have

$$V_c^d(f) \leq 1.$$

Choose a partition of $[a, b]$,

$$a = x_0 < x_1 < \cdots < x_N = b,$$

such that $x_k - x_{k-1} < \delta$ for all $k = 1, 2, \ldots, N$. Then

$$V_a^b(f) = \sum_{k=1}^{N} V_{x_{k-1}}^{x_k}(f) \leq N < \infty.$$

which shows that f is of bounded variation on $[a, b]$.   □

This proposition implies the existence of continuous functions which are not absolutely continuous; for example, $f(x) = x \sin(1/x)$ is such a function.

Since an absolutely continuous function is of bounded variation, we have the following result from Corollary 4.8.

5.5. Corollary. *If f is absolutely continuous on [a, b], then f is differentiable almost everywhere on [a, b]. Furthermore, the derivative f' is integrable on [a, b].*

An absolutely continuous function, being a function of bounded variation, can be represented as the difference of two monotone increasing functions. We have, however, a much stronger decomposition for an absolutely continuous function.

5.6. Proposition. *Every absolutely continuous function on $[a, b]$ is the difference of two absolutely continuous monotone increasing functions on $[a, b]$.*

Proof. By the proof of Theorem 4.5, we have

$$f(x) = V(x) - [V(x) - f(x)],$$

where $V(x) = V_a^x(f)$. Therefore, it suffices to show that V is absolutely continuous on $[a, b]$. Given any $\varepsilon > 0$, let η be such that $0 < \eta < \varepsilon$. By the absolute continuity of f, corresponding to η there is a $\delta > 0$ such that for any finite collection of mutually disjoint intervals $(a_k, b_k) \subset [a, b]$ of total length less than δ, we have

$$\sum_{k=1}^{n} |f(b_k) - f(a_k)| < \eta.$$

Then

$$\sum_{k=1}^{n} |V(b_k) - V(a_k)| = \sum_{k=1}^{n} V_{a_k}^{b_k}(f)$$

is the supremum of the sums

$$\sum_{k=1}^{n} \left(\sum_{j=1}^{n_k} |f(x_{k,j}) - f(x_{k,j-1})| \right),$$

where $a_k = x_{k,0} < x_{k,1} < \cdots < x_{k,n_k} = b_k$ is an arbitrary partition of $[a_k, b_k]$. Since the total length of the intervals $(x_{k,j-1}, x_{k,j})$ is clearly less than δ, the double sums above are less than η, since f is absolutely continuous. Therefore,

$$\sum_{k=1}^{n} |V(b_k) - V(a_k)| \le \eta < \varepsilon.$$

Thus V is absolutely continuous on $[a, b]$.   □

We devote the remaining part of this section to a characterization of absolutely continuous functions.

It is not true in general that the measurability of a set is invariant under a continuous function; that is, there is a continuous function which takes a measurable set into a nonmeasurable set.

5.7. Example. Let F be the Cantor ternary set, and let E be a generalized Cantor set which is not of measure zero (see Example 4.7, Chapter I). The intervals removed from $[0, 1]$ to construct the set F may be put into an obvious one–one correspondence with those intervals removed from $[0, 1]$ to construct the set E so as to preserve the natural order of the points. Let f

be the function on $[0, 1]\backslash F$ which maps each interval in $[0, 1]\backslash F$ onto the corresponding interval in $[0, 1]\backslash E$ linearly. This is explained best by the following figure:

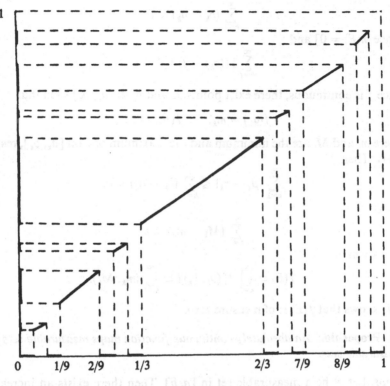

Figure 5.6

Extend the definition of f to the entire interval $[0, 1]$ in a natural way as described in the definition of the Lebesgue singular function (see Example 4.9). Then f is continuous on $[0, 1]$. Let A be a nonmeasurable subset of E. (Such a set exists by Proposition 4.6, Chapter I.) Then there exists a set $B \subset F$ such that

$$f(B) = A.$$

However, B is of measure zero; hence B is measurable. Therefore, the property of measurability is not invariant under a continuous function. □

We shall show that the measurability of a set is an invariant property under an absolutely continuous function.

5.8. Proposition. *An absolutely continuous function maps sets of measure zero into sets of measure zero.*

Proof. Let f be absolutely continuous on $[a, b]$ and let $E \subset [a, b]$ be of measure zero. We may assume that $E \subset (a, b)$. For an arbitrary $\varepsilon > 0$, we can find $\delta > 0$ and a sequence of pairwise disjoint intervals (a_k, b_k) covering E such that

$$\sum_{k=1}^{\infty} (b_k - a_k) < \delta$$

[since $m(E) = 0$] and

$$\sum_{k=1}^{\infty} |f(b_k) - f(a_k)| < \varepsilon.$$

Since f is continuous, there exist points α_k and β_k in $[a_k, b_k]$ such that

$$f(\alpha_k) = m_k, \qquad f(\beta_k) = M_k,$$

where m_k and M_k are the minimum and the maximum of f on $[a_k, b_k]$, respectively. Then

$$\sum_{k=1}^{\infty} |\beta_k - \alpha_k| \le \sum_{k=1}^{\infty} (b_k - a_k) < \delta.$$

Hence

$$\sum_{k=1}^{\infty} (M_k - m_k) < \varepsilon.$$

Also we have

$$f(E) \subset \bigcup_{k=1}^{\infty} f[(a_k, b_k)] \subset \bigcup_{k=1}^{\infty} (m_k, M_k).$$

This shows that $f(E)$ is of measure zero. $\square$

5.9. Proposition. *An absolutely continuous function maps measurable sets into measurable sets.*

Proof. Let A be a measurable set in $[a, b]$. Then there exists an increasing sequence (F_n) of closed sets contained in A such that

$$m(A) = \lim m(F_n)$$

by virtue of Proposition 5.5, Chapter III. Therefore, we can write

$$A = \left(\bigcup_{n=1}^{\infty} F_n \right) \cup N,$$

where N is of measure zero; hence

$$f(A) = f\left(\bigcup_{n=1}^{\infty} F_n \right) \cup f(N) = \left[\bigcup_{n=1}^{\infty} f(F_n) \right] \cup f(N).$$

Since $f(N)$ is of measure zero by Proposition 5.8, it remains to show that each $f(F_n)$ is measurable. Since f is continuous and the image of a compact set under a continuous function is also compact, $f(F_n)$ is compact; thus it is measurable. $\square$

Analyzing the preceding proof and the last six lines of Example 5.7, we can make the following assertion:

5.10. Corollary. *A continuous function maps measurable sets into measurable sets if and only if it maps sets of measure zero into sets of measure zero.*

We have shown that an absolutely continuous function is of bounded variation and maps measurable sets into measurable sets. In fact, these two properties characterize the class of absolutely continuous functions. The following theorem is due to S. Banach (1925):

5.11. Theorem (Banach, 1925). *A continuous function of bounded variation is absolutely continuous if and only if it maps measurable sets into measurable sets.*

The proof is not obvious. We refer the interested reader to Natanson (1955), pp. 250–252, or J. von Neumann (1950), pp. 79–81.

EXERCISES 5

A. Show that the two definitions of absolute continuity (i.e., Definition 5.1 and the one following Definition 5.1) are equivalent.

B. Show that the Lebesgue singular function described in §4 is not absolutely continuous. *Hint:* The Cantor ternary set F is compact and is of measure zero.

C. Prove Proposition 5.2.

D. Show that a function $f:[a, b] \to \mathbf{R}$ satisfies a Lipschitz condition if and only if for any $\varepsilon > 0$ there exists a $\delta > 0$ such that for any finite collection of intervals $(a_k, b_k) \subset [a, b]$, $k = 1, 2, \ldots, n$, with $\sum_{k=1}^{n} (b_k - a_k) < \delta$, we have

$$\sum_{k=1}^{n} |f(b_k) - f(a_k)| < \varepsilon.$$

[Notice that in Definition 5.1 we require that the intervals (a_k, b_k) are pairwise disjoint.]

E. Prove Proposition 5.3.

F. Let f be absolutely continuous on $[a, b]$ and $f([a, b]) = [c, d]$. If $g:[c, d] \to \mathbf{R}$ satisfies a Lipschitz condition, then show that the composite function $g(f(x))$ is absolutely continuous.

G. Show that the composite function of two absolutely continuous functions may not be absolutely continuous.

§6. The Fundamental Theorem of Calculus

We are now ready to prove the fundamental theorem of calculus for Lebesgue integrable functions.

6.1. The Fundamental Theorem I (Lebesgue, 1904). *If f is integrable on* $[a, b]$, *then the indefinite integral*

$$F(x) = \int_a^x f(t)\, dt$$

is absolutely continuous on $[a, b]$ *and*

$$F'(x) = f(x)$$

almost everywhere on $[a, b]$.

Proof. The absolute continuity of F follows from Proposition 1.13, Chapter IV. It remains to show that $F'(x) = f(x)$ for almost all x in $[a, b]$. It will be sufficient to show that the inequality

$$F'(x) \le f(x) \tag{*}$$

holds for almost all $x \in [a, b]$. If this is the case, changing f to $-f$ and F to $-F$ in (*), we obtain another inequality

$$-F'(x) \le -f(x)$$

or

$$F'(x) \ge f(x),$$

which holds for almost all $x \in [a, b]$; thus $F'(x) = f(x)$ almost everywhere. We shall prove (*) in the following lemma.   □

6.2. Lemma. *If f is integrable on* $[a, b]$ *and F is the indefinite integral of f, then* $F'(x) \le f(x)$ *for almost all* $x \in [a, b]$.

Proof. We show that the set

$$E = \{x \in [a, b] : F'(x) > f(x)\}$$

is of measure zero. We can write

$$E = \bigcap_{p, q} E_{pq},$$

where p and q are rationals with $p < q$ and

$$E_{pq} = \{x \in [a, b] : f(x) < p < q < F'(x)\}.$$

Therefore, it suffices to show that each E_{pq} is of measure zero.

We now prove that E_{pq} is of measure zero. By Proposition 1.13, Chapter IV, given any $\varepsilon > 0$ there is a $\delta > 0$ such that $m(A) < \delta$ implies

$$\left| \int_A f(t)\, dt \right| < \varepsilon. \tag{1}$$

Since f and F' are both measurable, the set E_{pq} is measurable; hence there is an open set G such that

$$E_{pq} \subset G, \quad m(G \setminus E_{pq}) < \delta,$$

by Proposition 5.2, Chapter III. Since G is an open set, we can decompose G as

$$G = \bigcup_{n=1}^{\infty} (a_n, b_n),$$

where (a_n, b_n) are pairwise disjoint. For each n we denote

$$G_n = E_{pq} \cap (a_n, b_n).$$

Then

$$G_n \subset \{x \in (a_n, b_n): F'(x) = D^+ F(x) > q\}.$$

We virtually repeat the proof of Lemma 3.5. Let $x_0 \in G_n$. Then there exists a point $\zeta > x_0$ such that

$$F(\zeta) - q\zeta > F(x_0) - qx_0.$$

Therefore, x_0 is a shadow point of the function $F(x) - qx$ with respect to the rising sun. This implies that G_n can be covered by a sequence of pairwise disjoint intervals $(a_{nk}, b_{nk}) \subset (a_n, b_n)$ such that

$$q(b_{nk} - a_{nk}) \le F(b_{nk}) - F(a_{nk}) = \int_{a_{nk}}^{b_{nk}} f(t)\, dt$$

by virtue of the Rising Sun Lemma. Therefore,

$$qm(G_n) \le \sum_{k=1}^{\infty} \int_{a_{nk}}^{b_{nk}} f(t)\, dt = \int_{S_n} f(t)\, dt,$$

where $S_n = \bigcup_{k=1}^{\infty} (a_{nk}, b_{nk})$. The S_n's are obviously pairwise disjoint. Let $S = \bigcup_{n=1}^{\infty} S_n$. Then it is clear that

$$E_{pq} \subset S \subset G, \quad m(S \setminus E_{pq}) < \delta.$$

Hence

$$\left| \int_{S \setminus E_{pq}} f(t)\, dt \right| < \varepsilon \tag{2}$$

by (1). Now

$$qm(E_{pq}) = q \sum_{n=1}^{\infty} m(G_n) \leq \sum_{n=1}^{\infty} \int_{S_n} f(t) \, dt$$

$$= \int_S f(t) \, dt = \int_{E_{pq}} f(t) \, dt + \int_{S \setminus E_{pq}} f(t) \, dt$$

$$\leq pm(E_{pq}) + \varepsilon$$

by the definition of E_{pq} and (2). Therefore,

$$m(E_{pq}) \leq \frac{\varepsilon}{q - p}$$

since ε is arbitrary, we conclude that E_{pq} is of measure zero.   □

6.3. Corollary. *If f is integrable on $[a, b]$, then there exists an absolutely continuous function F on $[a, b]$ such that*

$$f(x) = F'(x)$$

for almost all $x \in [a, b]$.

We need the following lemma in order to establish the fundamental relation:

$$\int_a^x f'(t) \, dt = f(x) - f(a).$$

6.4. Lemma. *If f is an absolutely continuous monotone increasing function on $[a, b]$ and $f'(x) = 0$ for almost all $x \in [a, \quad b]$, then f is a constant.*

Proof. Since f is continuous and monotone increasing, its range is the closed interval $[f(a), f(b)]$. To prove the lemma, it suffices to show that $[f(a), f(b)]$ is of measure zero; hence $f(x) = f(a)$ for all $x \in [a, b]$. Let $E = \{x \in [a, b] : f'(x) = 0\}$ and let $Z = [a, b] \setminus E$. Then, by hypothesis, we have $m(E) = b - a$ and $m(Z) = 0$. Clearly, $[f(a), f(b)] = f(E) \cup f(Z)$. Since f is absolutely continuous, $m(f(Z)) = 0$ by Proposition 5.8.

We must show that $f(E)$ is of measure zero. Let $x_0 \in E$. Since $f'(x_0) = 0$, we can find a point $\xi > x_0$ such that

$$\frac{f(\xi) - f(x_0)}{\xi - x_0} < \varepsilon,$$

or

$$\varepsilon\xi - f(\xi) > \varepsilon x_0 - f(x_0).$$

Therefore x_0 is a shadow point of the function $\varepsilon x - f(x)$ with respect to the rising sun. It follows from the Rising Sun Lemma that E is covered by countably many pairwise disjoint intervals (a_k, b_k) such that

$$\varepsilon a_k - f(a_k) \leq \varepsilon b_k - f(b_k),$$

or

$$f(b_k) - f(a_k) \leq \varepsilon(b_k - a_k).$$

Thus

$$\sum_{k=1}^{\infty} [f(b_k) - f(a_k)] \leq \varepsilon(b - a).$$

This means that $f(E)$ is covered by countably many intervals whose total length is arbitrarily small. Therefore, $f(E)$ is of measure zero. □

We observe that in the above proof we established the following assertion:

6.5. Corollary. *Let f be an absolutely continuous monotone increasing function on $[a, b]$. Then the set*

$$f(\{x \in [a, b]: f'(x) = 0\})$$

is of measure zero.

From this corollary we see that the Lebesgue singular function is not absolutely continuous.

We are now in a position to prove the fundamental theorem.

6.6. The Fundamental Theorem II (Lebesgue, 1904). *If f is absolutely continuous on $[a, b]$, then f' is integrable on $[a, b]$ and*

$$\int_a^x f'(t) \, dt = f(x) - f(a)$$

for all $x \in [a, b]$.

Proof. We need only prove the theorem for the case when f is monotone increasing on $[a, b]$ (why?). Since f is of bounded variation on $[a, b]$, f' is integrable on $[a, b]$ and

$$\int_a^x f'(t) \, dt \leq f(x) - f(a)$$

for all $x \in [a, b]$. Let

$$g(x) = f(x) - \int_a^x f'(t) \, dt.$$

Notice that $g(a) = f(a)$. Then g is absolutely continuous on $[a, b]$, since the difference of two absolutely continuous functions is also absolutely continuous. Moreover, g is monotone increasing since, if $a \leq x < y \leq b$, then

$$g(y) - g(x) = f(y) - f(x) - \int_x^y f'(t) \, dt \geq 0.$$

Furthermore, $g'(x) = 0$ almost everywhere by the Fundamental Theorem I.

Therefore g is a constant and is equal to $f(a)$. This proves that

$$\int_a^x f'(t)\, dt = f(x) - f(a).$$ □

6.7. Corollary. *Every absolutely continuous function f on $[a, b]$ can be represented in the form*

$$f(x) = \int_a^x g(t)\, dt + C,$$

where g is an integrable function on $[a, b]$ and C is a constant.

Proof. We have merely to define

$$g(x) = f'(x)$$

at the points in $[a, b]$ where f is differentiable and assign $g(x)$ arbitrary values at the other points of $[a, b]$. Then $C = f(a)$. □

Lemma 6.4 can be generalized for arbitrary absolutely continuous functions.

6.8. Corollary. *If f is absolutely continuous on $[a, b]$ and $f'(x) = 0$ for almost all $x \in [a, b]$, then f is a constant.*

Proof. Since

$$0 = \int_a^x f'(t)\, dt = f(x) - f(a),$$

$f(x) = f(a)$ for all $x \in [a, b]$. □

As an application of the Fundamental Theorems I and II, we have the following result.

6.9. Proposition (Integration by Parts). *Suppose F is absolutely continuous and g is integrable on $[a, b]$. Define $f(x) = F'(x)$ almost everywhere on $[a, b]$ and let $G(x) = \int_a^x g(t)\, dt + C$, where C is a constant. Then Fg and fG are integrable on $[a, b]$ and*

$$\int_a^b F(t)g(t)\, dt + \int_a^b f(t)G(t)\, dt = F(b)G(b) - F(a)G(a).$$

Proof. Since both F and G are absolutely continuous on $[a, b]$, so is FG, by Proposition 5.3. Hence $(FG)'$ is integrable and

$$\int_a^b (FG)'(t)\, dt = F(b)G(b) - F(a)G(a).$$

On the other hand,

$$(FG)'(t) = F(t)g(t) + f(t)G(t)$$

for almost all $t \in [a, b]$. Since F and G are bounded on $[a, b]$ and f and g are integrable, Fg and fG are integrable by Corollary 1.8, Chapter III. Therefore, we now get

$$\int_a^b F(t)g(t)\,dt + \int_a^b f(t)G(t)\,dt = F(b)G(b) - F(a)G(a). \qquad \square$$

We have discussed only the case where f is absolutely continuous in the fundamental relation between differentiation and integration. If we remove this restriction, the function f' need not be integrable (see Exercise 6A). The following problem arises: If f' is not integrable on $[a, b]$, how can f be reconstructed from f'? This problem involves a generalization of the integral concept which is outside the scope of this book. The Denjoy integral was invented just for the solution. For a thorough treatment the reader may consult S. Saks, *Theory of the Integral* (1937).

EXERCISES 6

A. Let

$$f(x) = \begin{cases} x^2 \sin \pi/x^2 & \text{if } x \neq 0, \\ 0 & \text{if } x = 0. \end{cases}$$

Show that f' exists on $(0, 1)$, but f' is not integrable on $[0, 1]$. *Hint:*

$$\int_0^1 \frac{1}{x}\left|\cos\frac{\pi}{x^2}\right|\,dx = \infty.$$

B. Let f be continuous and differentiable on $[a, b]$. Suppose that f' is bounded. Show that

$$\int_a^x f'(t)\,dt = f(x) - f(a)$$

for all $x \in [a, b]$.

C. Let f be integrable on $[a, b]$ and

$$\int_a^x f(t)\,dt = 0$$

for all $x \in [a, b]$. Show that $f = 0$ almost everywhere.

D. Using Exercise C and the Fundamental Theorem II, prove the Fundamental Theorem I.

E. Let f and g be absolutely continuous on $[a, b]$ such that

$$f'(x) = g'(x)$$

almost everywhere. Show that $f = g + c$ for some constant c.

For the remaining exercises we need the following concepts. By a *curve* we mean a pair of continuous functions

$$C: \begin{cases} x = x(t), \\ y = y(t), \end{cases}$$

where $t \in [a, b]$. A curve C is said to be *rectifiable* if

$$l(P) = \sum_{k=1}^{n} \sqrt{[x(t_k) - x(t_{k-1})]^2 + [y(t_k) - y(t_{k-1})]^2}$$

is bounded by a constant for every partition

$$P: \quad a = t_0 < t_1 < \cdots < t_n = b.$$

We define the *length* of the curve C as sup $l(P)$, where P runs over all possible partitions of $[a, b]$.

F. Show that the curve C is rectifiable if and only if both $x(t)$ and $y(t)$ are of bounded variation.

G. Show that if C is rectifiable, then

$$l'(t) = \sqrt{[x'(t)]^2 + [y'(t)]^2}$$

for almost all t in $[a, b]$, where $l(t)$ is the length of that part of the curve C obtained on $[a, t] \subset [a, b]$.

H. Show that $l(t)$ is absolutely continuous on $[a, b]$ if and only if both $x(t)$ and $y(t)$ are absolutely continuous on $[a, b]$.

I. If C is rectifiable, then show that

$$l(C) \geq \int_a^b \sqrt{[x'(t)]^2 + [y'(t)]^2} \, dt,$$

where $l(C)$ denotes the length of C. The equality holds if and only if $x(t)$ and $y(t)$ are both absolutely continuous.

J. Give an example of a curve C satisfying

$$l(C) > \int_a^b \sqrt{[x'(t)]^2 + [y'(t)]^2} \, dt.$$

K. Find the length of the curve $C: (x(t), y(t))$, where $x(t) = t$ and $y(t)$ is the Lebesgue singular function on $[0, 1]$.

CHAPTER VI

The L^p Spaces and the Riesz–Fischer Theorem

We now depart from the study of functions to the study of function spaces. So far our interest has been in developing the Lebesgue integral. The purpose of this new chapter is to relate the Lebesgue theory of integration to functional analysis. The theory of integration developed in this book enables us to introduce certain spaces of functions that have properties which are of great importance in analysis as well as mathematical physics, in particular, quantum mechanics. These are the so-called L^p spaces of measurable functions f such that $|f|^p$ is integrable. Aside from the intrinsic importance of these spaces, we also examine some applications of results in the previous chapters. One of the most important applications is to Fourier theory. As we remarked before, Fourier theory was a key motivation of the new theory of integration. We will present here the L^2 version of Fourier series, and in particular establish the Riesz–Fischer theorem which identifies the L^2 and l^2 spaces through Fourier series. We hope that this chapter will whet the reader's appetite for further study of abstract spaces such as Banach and Hilbert spaces.

§1. The L^p Spaces $(1 \le p < \infty)$

We have introduced L^1 space in §7, Chapter II. The space L^1 consists of the Lebesgue integrable functions f on $[a, b]$ with the norm

$$\|f\| = \int_a^b |f(x)|\, dx,$$

where two equivalent functions were considered identical. The domain on which such an integrable function is defined need not be the closed interval $[a, b]$. It can be any measurable set E. In this section we will generalize the concept of L^1 space and introduce the space L^p.

Let p be a positive real number, and let E be a measurable set in $\mathbb{R}$. We define the space $L^p(E)$ as the set of measurable functions on E such that $|f|^p$ is integrable on E; we will also identify two functions in $L^p(E)$ which are equivalent. Where no confusion will arise, we will denote $L^p(E)$ by L^p. For $p = 1$, the present definition reduces to our earlier definition of L^1.

1.1. Proposition. *The space L^p is a vector space over $\mathbb{R}$; i.e., if f and g are in L^p and α and β are real numbers, then $\alpha f + \beta g$ belongs to L^p.*

Proof. Clearly, if $f \in L^p$ and $\alpha \in \mathbb{R}$, then $\alpha f \in L^p$. Therefore, it suffices to show that $f + g$ lies in L^p whenever f and g are in L^p. But, in fact, we have

$$|f + g|^p \le (|f| + |g|)^p \le (2 \max\{|f|, |g|\})^p$$

$$\le 2^p \max\{|f|^p, |g|^p\} \le 2^p(|f|^p + |g|^p).$$

Since $|f + g|^p$ is measurable and $|f|^p + |g|^p$ is integrable, we conclude that $|f + g|^p$ is integrable. $\qquad\square$

In order to endow the space L^p with a norm, it is necessary to introduce some inequalities.

1.2. Young's Inequality. *Suppose that α is a real-valued function defined on $[0, \infty)$ such that:*

(a) $\alpha(0) = 0$;

(b) *α is continuous on $[0, \infty)$; and*

(c) *α is strictly increasing and unbounded on $[0, \infty)$.*

Under these conditions, α has an inverse function β which is defined on $[0, \infty)$ and which obeys (a), (b), *and* (c) *with α replaced by β.*

Then, for any $a \ge 0$ and $b \ge 0$, we have

$$ab \le \int_0^a \alpha(x)\, dx + \int_0^b \beta(y)\, dy$$

with equality holding if and only if $\alpha(a) = b$.

This inequality was considered by the English mathematician W.H. Young, "On classes of summable functions and their Fourier series" (1912). In the following we give a geometric proof. For an analytic proof we refer the reader to F. Cunningham, Jr. and N. Grosman, "On Young's inequality" (1971).

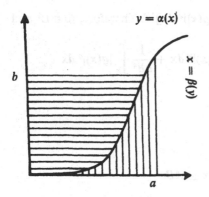 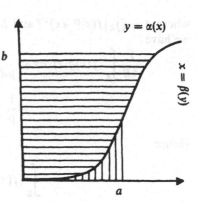

Figure 6.1

Proof. The result is obvious by considering Figure 6.1 and interpreting the integrals $\int_0^a \alpha(x)\,dx$ and $\int_0^b \beta(y)\,dy$ as the areas shaded with vertical lines and horizontal lines, respectively.   □

1.3. Corollary. *If $p > 1$, $a > 0$, and $b > 0$, then we have*

$$ab \le \frac{a^p}{p} + \frac{b^q}{q},$$

where q satisfies $1/p + 1/q = 1$. Equality holds if and if $a^p = b^q$.

Proof. Let $\alpha(x) = x^{p-1}$ and $\beta(y) = y^{1/(p-1)}$. Then α and β satisfy the conditions of Young's inequality. Therefore, by noting that $(p-1)q = p$, we have

$$ab \le \int_0^a x^{p-1}\,dx + \int_0^b y^{1/(p-1)}\,dy = \frac{a^p}{p} + \frac{b^q}{q}.$$

The assertion about equality is left for the reader to verify.   □

1.4. Hölder–Riesz Inequality. *Let $p > 1$ and $q > 1$ such that $1/p + 1/q = 1$. If $f \in L^p$ and $g \in L^q$, then $fg \in L^1$ and we have*

$$\int_E |f(x)g(x)|\,dx \le \left[\int_E |f(x)|^p\,dx\right]^{1/p}\left[\int_E |g(x)|^q\,dx\right]^{1/q}.$$

Proof. If f or g is zero almost everywhere, then the result is trivial. Otherwise, by Corollary 1.3, we have

$$\frac{|f(x)g(x)|}{AB} \le \frac{A^{-p}|f(x)|^p}{p} + \frac{B^{-q}|g(x)|^q}{q},$$

where $A = [\int_E |f(x)|^p \, dx]^{1/p}$ and $B = [\int_E |g(x)|^q \, dx]^{1/q}$. Therefore, $fg \in L^1$ and we have

$$\frac{1}{AB} \int_E |f(x)g(x)| \, dx \leq \frac{1}{pA^p} \int_E |f(x)|^p \, dx + \frac{1}{qB^q} \int_E |g(x)|^q \, dx$$

$$= \frac{1}{p} + \frac{1}{q} = 1.$$

Hence

$$\int_E |f(x)g(x)| \, dx \leq AB.$$

This proves the inequality.   □

For $p = q = 2$, the Hölder–Riesz inequality is called the *Cauchy–Bunyakovsky–Schwarz inequality*. Cauchy first proved the inequality for square summable sequences in his *Cours d'Analyse* (1821). If (s_n) and (t_n) are sequences such that $\sum_{n=1}^{\infty} s_n^2 < \infty$ and $\sum_{n=1}^{\infty} t_n^2 < \infty$, then

$$\sum_{n=1}^{\infty} |s_n t_n| \leq \left[\sum_{n=1}^{\infty} s_n^2 \right]^{1/2} \left[\sum_{n=1}^{\infty} t_n^2 \right]^{1/2}.$$

This inequality was generalized to integrals by the Russian mathematician Victor Bunyakovsky (1859). His contribution was overlooked by western authors, and later Hermann A. Schwarz discovered the inequality for integrals independently in 1885.

For general p and q, O. Hölder (1889) proved the inequality for sequences $(x_n), (y_n)$ such that $\sum_{n=1}^{\infty} |x_n|^p < \infty, \sum_{n=1}^{\infty} |y_n|^q < \infty$. The extension to integrals is due to F. Riesz (1910).

1.5. Minkowski–Riesz Inequality. *Let $p \geq 1$. If f and g are in L^p, then we have*

$$\left[\int_E |f(x) + g(x)|^p \, dx \right]^{1/p} \leq \left[\int_E |f(x)|^p \, dx \right]^{1/p} + \left[\int_E |g(x)|^p \, dx \right]^{1/p}.$$

Proof. For $p = 1$, the inequality is obtained by integrating the triangle inequality for real numbers. For $p > 1$, we proceed as follows:

$$\int_E |f(x) + g(x)|^p \, dx$$

$$\leq \int_E |f(x) + g(x)|^{p-1} |f(x)| \, dx + \int_E |f(x) + g(x)|^{p-1} |g(x)| \, dx.$$

Let $q > 0$ be such that $1/p + 1/q = 1$. Applying the Hölder–Riesz inequality to each of these last two integrals and noting that $(p - 1)q = p$, we have

$$\int_E |f(x) + g(x)|^p \, dx \leq M \left[\int_E |f(x) + g(x)|^{(p-1)q} \, dx \right]^{1/q}$$

$$= M \left[\int_E |f(x) + g(x)|^p \, dx \right]^{1/q}, \qquad (*)$$

where M denotes the right-hand side of the inequality we wish to prove. Now divide the extreme ends of the relation $(*)$ by

$$\left[\int_E |f(x) + g(x)|^p \, dx \right]^{1/q}$$

to obtain the desired result. $\square$

The German mathematician Hermann Minkowski (1864–1990) proved the preceding inequality for finite sums of numbers in his *Geometrie der Zahlen* (1896); then F. Riesz (1910) extended it to integrals of functions.

It should be noticed that the Hölder–Riesz inequality and the Minkowski–Riesz inequality do not hold for $0 < p < 1$ if $m(E) > 0$ (see Exercises 1B, 1C, and 1D).

We come now to a crowning achievement of F. Riesz, namely the discovery of the general L^p spaces $(1 \leq p < \infty)$ in his "Untersuchungen über Systeme integrierbarer Funktionen" (1910). Riesz's main tools in the study of the L^p spaces were the Hölder–Riesz and Minkowski–Riesz inequalities.

We now describe the exact sense in which the space L^p becomes a normed space (see §7, Chapter II, for the definition of normed spaces) for $p \geq 1$. As remarked before, we agree that $f = g$ means $f = g$ almost everywhere.

1.6. Proposition. *For $1 \leq p < \infty$, L^p is a normed space over $\mathbb{R}$ with respect to the norm*

$$\|f\|_p = \left[\int_E |f(x)|^p \, dx \right]^{1/p}.$$

Proof. We show that $\|f\|_p$ satisfies the following conditions:

(1) $\|f\|_p \geq 0$, $\|f\|_p = 0$ if and only if $f = 0$;
(2) $\|\alpha f\|_p = |\alpha| \, \|f\|_p$; and
(3) $\|f + g\|_p \leq \|f\|_p + \|g\|_p$.

Conditions (1) and (2) are trivial. Condition (3) is the Minkowski–Riesz inequality. Therefore, L^p is a normed space if $p \geq 1$. $\square$

The norm $\| \cdot \|_p$ will be called the L^p norm.

The reader may wonder why we restrict ourselves to $p \geq 1$. It turns out that L^p with $0 < p < 1$ is not really interesting (some would disagree!). In particular, the function $f \to \|f\|_p$ is not a norm if $m(E) > 0$ (see Exercise 1D).

We have the following theorem which is of vital importance in applications of Lebesgue integrals. (For $p = 2$, this theorem was proved independently by F. Riesz and E. Fischer in 1907. For an arbitrary p, $1 \le p < \infty$, it was shown by F. Riesz in 1910. However, the completeness of the space L^2 is so well recognized that the name "Riesz–Fischer theorem" is given to the following theorem.)

1.7. Riesz–Fischer Theorem. *For $1 \le p < \infty$, the space L^p is a Banach space.*

Proof. This proof is parallel to that of Theorem 7.3, Chapter II. Let (f_n) be a Cauchy sequence in L^p. Then, there is a natural number n_1 such that for all $n > n_1$ we have

$$\|f_n - f_{n_1}\|_p < \tfrac{1}{2}.$$

By induction, after finding $n_{k-1} > n_{k-2}$, we find $n_k > n_{k-1}$ such that for all $n > n_k$ we have

$$\|f_n - f_{n_k}\|_p < \frac{1}{2^k}.$$

Then (f_{n_k}) is a subsequence of (f_n) which satisfies

$$\|f_{n_{k+1}} - f_{n_k}\|_p < \frac{1}{2^k}, \qquad k = 1, 2, \ldots,$$

or

$$\|f_{n_1}\|_p + \sum_{k=1}^{\infty} \|f_{n_{k+1}} - f_{n_k}\|_p = A < \infty.$$

Let

$$g_k = |f_{n_1}| + |f_{n_2} - f_{n_1}| + \cdots + |f_{n_{k+1}} - f_{n_k}|, \qquad k = 1, 2, \ldots.$$

Then, by the Minkowski–Riesz inequality,

$$\int_E g_k^p(x)\, dx = \int_E (|f_{n_1}| + |f_{n_2} - f_{n_1}| + \cdots + |f_{n_{k+1}} - f_{n_k}|)^p\, dx$$

$$\le \left(\|f_{n_1}\|_p + \sum_{i=1}^{k} \|f_{n_{i+1}} - f_{n_i}\|_p \right)^p \le A^p < \infty.$$

Let $g = \lim g_k$. Then $g^p = \lim g_k^p$. By the Beppo Levi Theorem 1.5, Chapter IV, we have

$$\int_E g^p(x)\, dx = \lim \int_E g_k^p(x)\, dx < \infty.$$

This shows that g is in L^p, and hence

$$\int_E \left(|f_{n_1}| + \sum_{k=1}^{\infty} |f_{n_{k+1}} - f_{n_k}| \right)^p dx < \infty.$$

This implies that

$$f_{n_1}(x) + \sum_{k=1}^{\infty} [f_{n_{k+1}}(x) - f_{n_k}(x)]$$

converges almost everywhere to a function f in L^p.

It remains to prove that $\|f_{n_k} - f\|_p \to 0$ as $k \to \infty$. We first notice that

$$f(x) - f_{n_j}(x) = \sum_{k=j}^{\infty} [f_{n_{k+1}}(x) - f_{n_k}(x)].$$

It follows that

$$\|f - f_{n_j}\|_p \leq \sum_{k=j}^{\infty} \|f_{n_{k+1}} - f_{n_k}\|_p < \sum_{k=j}^{\infty} \frac{1}{2^k} = \frac{1}{2^{j-1}}.$$

Therefore, $\|f - f_{n_j}\|_p \to 0$ as $j \to \infty$. Now

$$\|f_n - f\|_p \leq \|f_n - f_{n_k}\|_p + \|f_{n_k} - f\|_p,$$

where $\|f_n - f_{n_k}\|_p \to 0$ as $n \to \infty$ and $k \to \infty$ and hence $\|f_n - f\|_p \to 0$ as $n \to \infty$. This shows that the Cauchy sequence (f_n) converges to f in L^p. $\square$

The preceding proof contains a result which is interesting enough to be stated separately.

1.8. Proposition. *If $1 \leq p < \infty$ and if (f_n) is a Cauchy sequence in L^p with limit f, then (f_n) has a subsequence which converges pointwise almost everywhere to f.*

Before closing this section we would like to establish a few more simple properties of the L^p spaces using the Hölder–Riesz inequality.

1.9. Proposition. *If $m(E) < \infty$ and $0 < q < p$, then $L^p \subset L^q$ and there exists a constant $C > 0$ such that $\|f\|_q \leq C\|f\|_p$ for all $f \in L^p$.*

Proof. For any $f \in L^p$, we have

$$\int_E |f(x)|^p \, dx < \infty.$$

It is evident that f^q belongs to L^r where $r = p/q > 1$. Let s be such that $1/r + 1/s = 1$. Then

$$\int_E |f(x)|^q \, dx = \int_E |f(x)|^q \cdot 1 \, dx$$

$$\leq \left[\int_E |f(x)|^{qr} \, dx \right]^{1/r} \left[\int_E 1^s \, dx \right]^{1/s}$$

$$= \left[\int_E |f(x)|^p \, dx \right]^{q/p} [m(E)]^{1/s}.$$

Therefore, $f \in L^q$ and

$$\|f\|_q \leq C\|f\|_p$$

where $C = [m(E)]^{1/sq} > 0$. $\square$

If $m(E) = \infty$, the preceding property does not hold. A counterexample can be found easily.

1.10. Proposition. *Let* $0 < q < p < \infty$. *If* $f \in L^p \cap L^q$ *then* $f \in L^r$ *for all* $q < r < p$.

Proof. Since $q < r < p$, there is a number t, $0 < t < 1$, such that $r = tq + (1 - t)p$. Note that $|f|^{tq} \in L^{1/t}$ and $|f|^{(1-t)p} \in L^{1/(1-t)}$. Hence, by the Hölder–Riesz inequality, we have $|f|^r = |f|^{tq}|f|^{(1-t)p} \in L^1$. $\square$

The Hölder–Riesz inequality implies that if $f \in L^p$ and $g \in L^q$ ($1/p + 1/q = 1$), then $fg \in L^1$. But it is not in general true that the product of two integrable functions is also integrable. In fact, we have the following proposition:

1.11. Proposition. *Let* f *be integrable on* E *such that* f *is not equivalent to any bounded function. Then there exists an integrable function* g *on* E *such that* fg *is not integrable on* E.

Proof. For each natural number n, consider the following set:

$$E_n = \{x \in E: n \leq |f(x)| < n + 1\}.$$

Since f is not equivalent to a bounded function, $m(E_n) < \infty$ for an infinite number of values of n. Therefore, there exists a sequence (a_n) in $\mathbb{N}$ such that $a_n \geq n$, $a_n \uparrow \infty$, and $0 < m(A_n) < \infty$, where

$$A_n = \{x \in E: a_n \leq |f(x)| < a_n + 1\}.$$

Denote $c_n = m(A_n)$. Then

$$\sum_{n=1}^{\infty} a_n c_n \leq \int_E |f(x)|\, dx < \infty.$$

Define $g: E \to \mathbb{R}$ by

$$g(x) = \begin{cases} 1/na_n c_n & \text{if } x \in A_n, \\ 0 & \text{if } x \notin \bigcup_{n=1}^{\infty} A_n. \end{cases}$$

Then

$$\int_E g(x)\, dx = \sum_{n=1}^{\infty} \frac{1}{na_n} \leq \sum_{n=1}^{\infty} \frac{1}{n^2} < \infty$$

showing that g is integrable on E. However,

$$\int_{A_n} |f(x)g(x)| \, dx \geq \frac{1}{n}$$

and $\sum_{n=1}^{\infty} (1/n) = \infty$. Therefore, fg is not integrable on E. $\qquad\square$

EXERCISES 1

A. Equality in the Hölder–Riesz inequality holds if and only if there exist non-negative numbers α and β such that $\alpha |f(x)|^p = \beta |g(x)|^q$ almost everywhere.

B. **Hölder–Riesz Inequality for $0 < p < 1$.** *Let $0 < p < 1$ and let $f \in L^p$ and $g \in L^q$ such that $f \geq 0$ and $g > 0$ a.e. Then we have*

$$\int_E f(x)g(x) \, dx \geq \left[\int_E f^p(x) \, dx \right]^{1/p} \left[\int_E g^q(x) \, dx \right]^{1/q},$$

provided

$$\int_E g^q(x) \, dx \neq 0, \qquad \text{since} \quad q < 0.$$

C. **Minkowski–Riesz Inequality for $0 < p < 1$.** *Let $0 < p < 1$ and $f, g \in L^p$ be such that $f \geq 0$, $g \geq 0$. Then*

$$\|f + g\|_p \geq \|f\|_p + \|g\|_p.$$

D. Let $0 < p < 1$ and $m(E) > 0$. Then there exist f and g in L^p such that

$$\|f + g\|_p > \|f\|_p + \|g\|_p.$$

Hint: Find two disjoint subsets A and B of positive measure in E. Let $f = \alpha \chi_A$ and $g = \beta \chi_B$, where $\alpha > 0$, $\beta > 0$. Compute the norms $\|f\|_p$, $\|g\|_p$, $\|f + g\|_p$. Now adjust α and β to suit the problem.

E. Let $0 < q < p$. Find a function f in $L^p(\mathbb{R})$ which is not in $L^q(\mathbb{R})$.

F. Let (f_n) be a sequence in $L^p(E)$ which converges to a function f in $L^p(E)$ with respect to the L^p norm. Suppose that g is a pointwise limit of the sequence (f_n). Show that $f = g$ almost everywhere.

§2. Approximations by Continuous Functions

In §7, Chapter II, it was proved that the space $C[a, b]$ of continuous real-valued functions on $[a, b]$ is dense in $L^1[a, b]$ with respect to the L^1 norm. We now want to generalize this result, first from $[a, b]$ to any measurable set and then from $L^1(E)$ to $L^p(E)$.

If $m(E) = \infty$, it is not true in general that a continuous function on E is integrable on E; e.g., a constant function on $\mathbb{R}$ is not integrable. Therefore the space $C(E)$ of continuous functions on E is not, in general, a subset of $L^1(E)$.

This suggests that our generalization should be more restrictive. We first consider the case $E = \mathbb{R}$.

2.1. Proposition. *For $1 \leq p < \infty$, the set of all step functions on $\mathbb{R}$ is dense in $L^p(\mathbb{R})$; i.e., if $f \in L^p(\mathbb{R})$ and $\varepsilon > 0$, then there exists a step function φ on $\mathbb{R}$ such that $\|f - \varphi\|_p < \varepsilon$.*

Proof. This proof consists of two steps. First we consider $f \geq 0$. Let $\varepsilon > 0$. Then we can find a step function $\varphi \geq 0$ such that

$$\|f^p - \varphi^p\|_1 < \varepsilon^p.$$

(Why?) Using the inequality

$$|a - b|^p \leq |a^p - b^p|$$

which is valid for $a \geq 0$, $b \geq 0$, and $p \geq 1$, we obtain

$$\int_{\mathbb{R}} |f(x) - \varphi(x)|^p \, dx \leq \int_{\mathbb{R}} |f^p(x) - \varphi^p(x)| \, dx < \varepsilon^p,$$

or

$$\|f - \varphi\|_p < \varepsilon.$$

The general case follows directly from this by considering the decomposition $f = f^+ - f^-$. $\qquad\qquad\qquad\qquad\qquad\qquad\qquad\qquad\qquad\square$

To state the main result of this section we need to introduce a new class of functions defined on $\mathbb{R}$.

2.2. Definition. A continuous function $f: \mathbb{R} \to \mathbb{R}$ is said to have *compact support* if there is a compact set $K \subset \mathbb{R}$ such that $x \notin K$ implies $f(x) = 0$. The class of all such functions on $\mathbb{R}$ will be denoted by $C_c(\mathbb{R})$.

Since a compact set in $\mathbb{R}$ is bounded, if $f \in C_c(\mathbb{R})$, then f takes the value 0 outside a closed interval, and hence it is clear that $C_c(\mathbb{R}) \subset L^p(\mathbb{R})$.

Let E be a measurable subset of $\mathbb{R}$ and $f \in C_c(\mathbb{R})$. Consider the restriction f_E of f to E, defined by

$$f_E = \chi_E f.$$

Then f_E is continuous and integrable on E. Moreover, $f_E \in L^p(E)$. We denote

$$C_c(E) = \{f_E : f \in C_c(\mathbb{R})\}.$$

It is evident that $C_c(E) \subset C(E) \cap L^p(E)$. It is simple to show that if E is compact, then $C_c(E) = C(E)$ (see Exercise 2A).

2.3. Proposition. *For $1 \leq p < \infty$, $C_c(E)$ is dense in $L^p(E)$.*

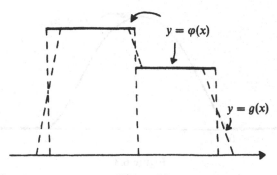

Figure 6.2

Proof. Let $f \in L^p(E)$. Define a function $\tilde{f}: \mathbb{R} \to \mathbb{R}$ by

$$\tilde{f}(x) = \begin{cases} f(x) & \text{if } x \in E, \\ 0 & \text{if } x \notin E. \end{cases}$$

Then $\tilde{f} \in L^p(\mathbb{R})$. Therefore, for any $\varepsilon > 0$ there is a step function φ on $\mathbb{R}$ such that

$$\int_E |f(x) - \varphi_E(x)|^p \, dx \le \int_{\mathbb{R}} |\tilde{f}(x) - \varphi(x)|^p \, dx < \frac{\varepsilon^p}{2^p}. \tag{*}$$

On the other hand, there exists a $g \in C_c(\mathbb{R})$ such that

$$\int_{\mathbb{R}} |g(x) - \varphi(x)|^p \, dx < \frac{\varepsilon^p}{2^p}. \tag{**}$$

This can easily be seen by considering Figure 6.2. Combining (*) and (**), we have

$$\left[\int_E |f(x) - g_E(x)|^p \, dx \right]^{1/p} \le \left[\int_{\mathbb{R}} |\tilde{f}(x) - g(x)|^p \, dx \right]^{1/p}$$

$$\le \left[\int_{\mathbb{R}} |\tilde{f}(x) - \varphi(x)|^p \, dx \right]^{1/p} + \left[\int_{\mathbb{R}} |\varphi(x) - g(x)|^p \, dx \right]^{1/p} < \varepsilon,$$

which shows that $C_c(E)$ is dense in $L^p(E)$. $\square$

2.4. Corollary. *For $1 \le p < \infty$, $C(E) \cap L^p(E)$ is dense in $L^p(E)$.*

We can do even better than Proposition 2.3 by smoothing out sharp corners of the graph of a continuous function which approximates an L^p function. More precisely, an L^p function can be approximated by an indefinitely differentiable continuous function with compact support.

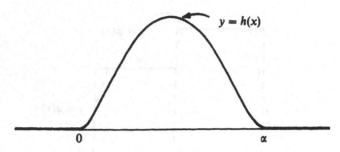

Figure 6.3

We let $C_c^\infty(\mathbb{R})$ be the space of infinitely differentiable functions on $\mathbb{R}$ with compact support. To approximate L^p functions by $C_c^\infty(\mathbb{R})$ functions, we now construct some basic functions in $C_c^\infty(\mathbb{R})$ which are useful in many areas of analysis.

Let α be a positive real number. The function

$$h(t) = \begin{cases} \exp[-1/t(\alpha - t)] & \text{if } 0 < t < \alpha, \\ 0 & \text{if } t \leq 0 \text{ or } t \geq \alpha, \end{cases}$$

is a bell-shaped function which is infinitely differentiable and posesses compact support (see Figure 6.3).

Define a function $g: \mathbb{R} \to \mathbb{R}$ by

$$g(x) = \int_{-\infty}^{x} h(t)\, dt.$$

Then g starts from 0 and climbs between 0 and α to a constant value (see Figure 6.4). Multiplying by a positive number, we can assume that the maximum value is equal to any preassigned positive number.

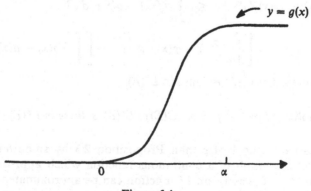

Figure 6.4

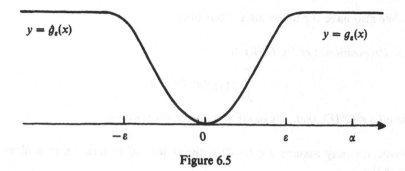

Figure 6.5

Let $\varepsilon > 0$. Consider the function $g_\varepsilon: \mathbb{R} \to \mathbb{R}$ defined by $g_\varepsilon(x) = g(\alpha x/\varepsilon)$. Then $g_\varepsilon(x) = g(\alpha)$ if $x \geq \varepsilon$ (see Figure 6.5). Let $\theta_\varepsilon(x) = g_\varepsilon(-x)$. We now define a function in $C_c^\infty(\mathbb{R})$. For the closed interval $[a, b]$ and $\varepsilon > 0$, let

$$p_\varepsilon(x) = \begin{cases} g_\varepsilon(x - a + \varepsilon) & \text{if } x \leq (a + b)/2, \\ g_\varepsilon(x - b - \varepsilon) & \text{if } x \geq (a + b)/2. \end{cases}$$

Then $p_\varepsilon \in C_c^\infty(\mathbb{R})$ (see Figure 6.6). It is clear that the characteristic function $\chi_{[a,b]}$ of $[a, b]$ can be approximated by rp_ε for some $r > 0$ and $\varepsilon > 0$ in the sense that

$$\|rp_{\varepsilon/2} - \chi_{[a,b]}\|_p < \varepsilon.$$

Let $C_c^\infty(E) = \{f_E: f \in C_c^\infty(\mathbb{R})\}$. Then we have the following proposition:

2.5. Proposition. *For* $1 \leq p < \infty$, $C_c^\infty(E)$ *is dense in* $L^p(E)$.

Proof. We may restrict ourselves to $E = \mathbb{R}$. We know that the step functions on $\mathbb{R}$ are dense in $L^p(\mathbb{R})$. On the other hand, the characteristic function of an interval can be approximated by C_c^∞ functions, as we have seen above. Therefore, the assertion follows at once.   □

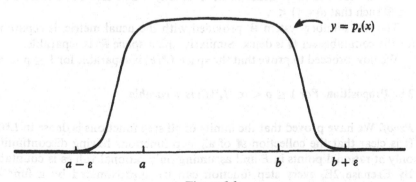

Figure 6.6

We also have the following proposition:

2.6. Proposition. *Let $f \in L^1(E)$. If*

$$\int_E f(x)g(x)\, dx = 0$$

for all $g \in C_c^\infty(E)$, then f is equal to 0 almost everywhere.

Proof. We may assume $f \in L^1(\mathbb{R})$ without loss of generality. It suffices to show that

$$\int_A f(x)\, dx = 0$$

for all bounded measurable subsets A of $\mathbb{R}$ (see Exercise 2B). Since $\chi_A \in L^1(\mathbb{R})$, we can find a sequence (g_n) in $C_c^\infty(\mathbb{R})$ which tends almost everywhere to χ_A and is bounded by a constant M. Then $(g_n f)$ converges to $f_A = \chi_A f$ almost everywhere; each $g_n f$ is in $L^1(\mathbb{R})$. By the Lebesgue Dominated Convergence Theorem, we conclude that $(g_n f)$ converges to f_A with respect to the L^1 norm, whence

$$\int_\mathbb{R} g_n(x)f(x)\, dx \to \int_\mathbb{R} f_A(x)\, dx.$$

This proves that $\int_A f(x)\, dx = \int_\mathbb{R} f_A(x)\, dx = 0$. $\square$

Finally we introduce the concept of separability of a metric space and prove that $L^p(E)$ is separable if $1 \le p < \infty$.

2.7. Definition. A metric space M is said to be *separable* if it posesses a dense subset $\mathscr{D}$ consisting of a countable number of elements.

Recall that a set $\mathscr{D}$ is dense in M if for any $x \in M$ and $\varepsilon > 0$ there exists $y \in \mathscr{D}$ such that $d(x, y) < \varepsilon$.

The real number system $\mathbb{R}$, provided with the usual metric, is separable, for the countable set $\mathbb{Q}$ is dense. Similarly, the n-space $\mathbb{R}^n$ is separable.

We now proceed to prove that the space $L^p(E)$ is separable for $1 \le p < \infty$.

2.8. Proposition. *For $1 \le p < \infty$, $L^p(E)$ is separable.*

Proof. We have proved that the family of all step functions is dense in $L^p(E)$. It is clear that the collection $\mathscr{R}$ of all step functions having discontinuities only at rational points in E and assuming only rational values is countable. By Exercise 2E, every step function can be approximated by a function in $\mathscr{R}$ with respect to the L^1 norm. Therefore the collection $\mathscr{R}$ is dense in $L^p(E)$. $\square$

EXERCISES 2

A. Show that $C_c[a, b] = C[a, b]$. More generally, show that $C_c(E) = C(E)$ if E is compact.

B. Let $f \in L^1(\mathbb{R})$ and suppose that

$$\int_A f(x)\, dx = 0$$

for all bounded measurable subsets A of $\mathbb{R}$. Show that f is equal to 0 almost everywhere.

C. Let $f \in L^2(\mathbb{R})$ and suppose that

$$\int_\mathbb{R} f(x)g(x)\, dx = 0$$

for all $g \in C_c^\infty(\mathbb{R})$. Show that f is equal to 0 almost everywhere. *Hint:* Proposition 2.5.

D. Let M be a metric space and let A be a dense subset of M. Suppose that A is separable. Show that M is separable.

E. Let φ be a step function on $\mathbb{R}$, with $\varepsilon > 0$. Show that there is a step function ψ having discontinuities only at rational points and assuming only rational values such that $\|\varphi - \psi\|_p < \varepsilon$.

§3. The Space L^∞

In the preceding sections we have studied the L^p spaces for $1 \le p < \infty$. It is natural to question whether there is an interpretation of L^p as $p \to \infty$. This will be answered by introducing a new class of functions.

3.1. Definition. Let E be a measurable set. We define $L^\infty(E)$ to be the space of all measurable functions on E which are bounded almost everywhere on E, i.e.,

$$m(\{x \in E: |f(x)| > \alpha\}) = 0$$

for some real number $\alpha > 0$.

It is easy to see that $L^\infty(E)$ is a vector space over the field $\mathbb{R}$. We shall endow $L^\infty(E)$ with a norm $\|\cdot\|_\infty$ defined by

$$\|f\|_\infty = \inf\{\alpha: m(\{x \in E: |f(x)| > \alpha\}) = 0\}.$$

This norm $\|f\|_\infty$ is sometimes called the *essential supremum* of f.

It can of course happen that $\|f\|_\infty < \sup_{x \in E} f(x)$, and we could even have $\|f\|_\infty < \infty$ and $\sup_{x \in E} |f(x)| = \infty$ (why?).

Let $\beta = \|f\|_\infty$. Since

$$\{x \in E: |f(x)| > \beta\} = \bigcup_{n=1}^\infty \left\{x \in E: |f(x)| > \beta + \frac{1}{n}\right\},$$

and since the union of a countable collection of sets of measure zero has measure zero, we see that

$$m(\{x \in E: |f(x)| > \|f\|_\infty\}) = 0,$$

i.e., $|f(x)| \le \|f\|_\infty$ for almost all $x \in E$.

The essential supremum $\|\cdot\|_\infty$ satisfies the conditions for a norm:

(1) $\|f\|_\infty \ge 0$;
(2) $\|f\|_\infty = 0$ if and only if $f = 0$ (almost everywhere);
(3) $\|\alpha f\|_\infty = |\alpha| \|f\|_\infty$ for $\alpha \in \mathbb{R}$; and
(4) $\|f + g\|_\infty \le \|f\|_\infty + \|g\|_\infty$.

These properties are very elementary, and we ask the reader to verify them.

Before considering more pertinent properties of this normed space $L^\infty(E)$, we shall justify our notations L^∞, $\|f\|_\infty$ for the sake of motivation. This is done in the following proposition:

3.2. Proposition. *Let E be a set of finite measure. Then*

$$L^\infty(E) \subset L^p(E)$$

for all p, $1 \le p < \infty$. Furthermore, if $f \in L^\infty(E)$, then

$$\|f\|_\infty = \lim_{p \to \infty} \|f\|_p.$$

Notice that we have never defined the notation $\lim_{p \to \infty} \|f\|_p$, where $p \in [1, \infty)$. By $\|f\|_\infty = \lim_{p \to \infty} \|f\|_p$ we mean that for any sequence (a_n) in $[0, \infty)$ with $a_n \uparrow \infty$, we have

$$\lim_{n \to \infty} \|f\|_{a_n} = \|f\|_\infty.$$

We recall here that a sequence (a_n) of real numbers converges if and only if

$$-\infty < \lim \inf a_n = \lim \sup a_n < \infty.$$

In this case, $\lim a_n = \lim \sup a_n$ (see Theorem 6.7, Chapter II).

Proof of Proposition. Let $f \in L^\infty(E)$ and $A = \|f\|_\infty$. Then $|f(x)|^p \le A^p$ for almost all $x \in E$. Therefore, $f \in L^p(E)$ and

$$\int_E |f(x)|^p \, dx \le A^p m(E),$$

or

$$\|f\|_p \le A[m(E)]^{1/p}.$$

Since $[m(E)]^{1/p} \to 1$ as $p \to \infty$, we have

$$\limsup \|f\|_p \leq A$$

[as remarked before, where p runs through every sequence (a_n), $a_n \uparrow \infty$]. On the other hand, suppose that $|f(x)| \geq B$ on a set F of positive measure. Then

$$B[m(F)]^{1/p} \leq \|f\|_p.$$

This shows that

$$B \leq \liminf \|f\|_p$$

and hence

$$\sup\{B: m(\{x \in E: |f(x)| \geq B\}) \neq 0\} \leq \liminf \|f\|_p.$$

But the left-hand side of the preceding inequality is equal to $A = \|f\|_\infty$ (Exercise 3A). Therefore,

$$A \leq \liminf \|f\|_p \leq \limsup \|f\|_p \leq A,$$

which proves that $A = \lim_{p \to \infty} \|f\|_p$. $\qquad\square$

Proposition 3.2 shows that $L^\infty(E) \subset \bigcap_{p \geq 1} L^p(E)$, and the norm on $L^\infty(E)$ is equal to the limit of $\|f\|_p$ as $p \to \infty$ provided $m(E) < \infty$.

3.3. Proposition. *The space $L^\infty(E)$ is a Banach space.*

Proof. Suppose that (f_n) is a Cauchy sequence in $L^\infty(E)$, and let

$$A_k = \{x \in E: |f_k(x)| > \|f_k\|_\infty\},$$

$$B_{m,n} = \{x \in E: |f_m(x) - f_n(x)| > \|f_m - f_n\|_\infty\},$$

$$F = \bigcup_{k=1}^\infty A_k \cup \bigcup_{m,n=1}^\infty B_{m,n}.$$

Then $m(F) = 0$. For each $x \in E\backslash F$, the sequence $(f_n(x))$ is a Cauchy sequence in $\mathbb{R}$ since

$$|f_m(x) - f_n(x)| \leq \|f_m - f_n\|_\infty.$$

Therefore it converges to a real number. Let $f: E \to \mathbb{R}$ be defined by

$$f(x) = \begin{cases} \lim f_n(x) & \text{if } x \in E\backslash F, \\ 0 & \text{if } x \in F. \end{cases}$$

Then it is clear that f is measurable and bounded, and hence $f \in L^\infty(E)$. We now show that (f_n) converges to f in the sense that $\|f_n - f\|_\infty \to 0$ as $n \to \infty$. Notice that $f_n \to f$ uniformly on $E\backslash F$, and $f_n - f$ is bounded on $E\backslash F$. Therefore,

$$\|f_n - f\|_\infty \leq \sup\{|f_n(x) - f(x)|: x \in E\backslash F\} \to 0$$

as $n \to \infty$. Therefore, $L^\infty(E)$ is complete. $\qquad\square$

In §2, we have shown that $C_c^\infty(E)$ is dense in $L^p(E)$ if $1 \leq p < \infty$. The case $p = \infty$ differs from the case $1 \leq p < \infty$. Surprisingly, no family of continuous functions is dense in $L^\infty(E)$. For simplicity's sake we will consider a bounded or unbounded interval (a, b) instead of a general set of positive measure in the following argument:

3.4. Proposition. $L^\infty(a, b)$ *is not separable.*

Proof. For each real number c, $a < c < b$, we define f_c to be the characteristic function of the interval (a, c). Then if $c \neq d$, we have $\|f_c - f_d\|_\infty = 1$. Suppose that a set $\mathscr{F}$ is dense in $L^\infty(a, b)$. Then for each c, $a < c < b$, there is a function g_c in $\mathscr{F}$ such that

$$\|f_c - g_c\|_\infty < \tfrac{1}{2}.$$

Therefore $\mathscr{F}$ is necessarily uncountable. This shows that $L^\infty(a, b)$ is not separable. □

3.5. Proposition. *No family of continuous functions on (a, b) is dense in* $L^\infty(a, b)$.

Proof. Let f_c be as described in the preceding proof. If f is a continuous function, then it is easy to show that $\|f_c - f\| \geq \tfrac{1}{2}$ (see Exercise 3E). This shows that no family of continuous functions on (a, b) is dense in $L^\infty(a, b)$. □

However, the following proposition holds; we leave the proof to the reader (see Exercise 3F):

3.6. Proposition. *The family of step functions on (a, b) is dense in* $L^\infty(a, b)$.

EXERCISES 3

A. Let $f \in L^\infty(E)$. Show that

$$\|f\|_\infty = \sup\{B : m\{x \in E : |f(x)| \geq B\} \neq 0\}.$$

B. Let $f \in L^\infty(E)$. Show that

$$\|f\|_\infty = \inf\left\{\sup_{x \in E} |g(x)| : g \sim f\right\}.$$

where $g \sim f$ means that $f = g$ almost everywhere.

C. Show that $\|\cdot\|_\infty$ is a norm on $L^\infty(E)$.

D. Let $f \in L^1(E)$ and $g \in L^\infty(E)$. S owhhat

$$fg \in L^1 \quad \text{and} \quad \|fg\|_1 \leq \|f\|_1 \|g\|_\infty.$$

E. Let f be continuous on (a, b). For c, $a < c < b$, let f_c be the characteristic function on (a, c). Show that $\|f_c - f\|_\infty \ge \frac{1}{2}$.

F. Prove Proposition 3.6.

G. Let E be a set of positive measure. Prove that $L^\infty(E)$ is not separable.

H. Show that the space $C_c(\mathbb{R})$ of continuous functions with compact support is not complete with respect to the norm $\|f\|_\infty = \sup_{x \in \mathbb{R}} |f(x)|$. Is $L^\infty(\mathbb{R})$ the completion of $C_c(\mathbb{R})$? In not, describe the completion of $C_c(\mathbb{R})$.

§4. The l^p Spaces $(1 \le p \le \infty)$

The purpose of this section is to introduce L^p space techniques to the classical sequence spaces l^p. The arguments presented in this section are unnecessarily elaborate, for no reference to integration is needed. Direct proofs of major results here can be supplied easily by the reader.

The n-space $\mathbb{R}^n$ can be made into normed spaces in a variety of ways. If $f = (\xi_1, \xi_2, \ldots, \xi_n) \in \mathbb{R}^n$ and $p \ge 1$, we define

$$\|f\|_p = \left[\sum_{k=1}^n |\xi_k|^p \right]^{1/p}.$$

$\mathbb{R}^n$ with the norm $\|f\|_p$ will be denoted by $l^p(n)$. In particular, the n-dimensional Euclidean space $\mathbb{R}^n$ is equal to $l^2(n)$ (see §7, Chapter Zero).

We also define a norm on $\mathbb{R}^n$ by

$$\|f\|_\infty = \sup_{1 \le k \le n} |\xi_k|.$$

$\mathbb{R}^n$ with the norm $\|f\|_\infty$ will be denoted by $l^\infty(n)$.

Naturally, we can ask whether it is possible to consider spaces analogous to $l^p(n)$ and $l^\infty(n)$ for infinite sequences.

4.1. Definition. For $1 \le p < \infty$, the space l^p is defined as the collection of all sequences $f = (\xi_n)$ of numbers such that $\sum_{n=1}^\infty |\xi_n|^p < \infty$, i.e.,

$$l^p = \left\{ (\xi_n): \sum_{n=1}^\infty |\xi_n|^p < \infty \right\}.$$

The norm on l^p is defined by

$$\|f\|_p = \left[\sum_{n=1}^\infty |\xi_n|^p \right]^{1/p}.$$

We define l^∞ as the collection of all bounded sequences $f = (\xi_n)$, i.e.,

$$l^\infty = \left\{ (\xi_n): \sup_{n \in \mathbb{N}} |\xi_n| < \infty \right\}.$$

The norm on l^∞ is defined by

$$\|f\|_\infty = \sup_{n \in N} |\xi_n|.$$

Although we can show directly by an elementary method that l^p, $1 \leq p \leq \infty$, is a Banach space, we will prove this fact indirectly using the completeness of L^p, $1 \leq p \leq \infty$. To do so we identify each sequence $f \in l^p$ with an integrable function $\tilde{f}$ in $L^p[0, \infty)$ where

$$\tilde{f}(x) = \xi_n \quad \text{if} \quad n - 1 \leq x < n.$$

Then

$$\|\tilde{f}\|_p = \left[\sum_{n=1}^{\infty} |\xi_n|^p \right]^{1/p}, \quad 1 \leq p < \infty,$$

$$\|\tilde{f}\|_\infty = \sup_{n \in N} |\xi_n|, \quad p = \infty.$$

Therefore

$$\|\tilde{f}\|_p = \|f\|_p, \quad 1 \leq p \leq \infty. \tag{*}$$

Under this convention we can treat l^p as a vector subspace of $L^p[0, \infty)$ for all p, $1 \leq p \leq \infty$. By the equation (*), l^p is a normed space for $1 \leq p \leq \infty$. [In particular, $l^p(n)$ is a normed space.]

4.2. Proposition. *For $1 \leq p \leq \infty$, l^p is a Banach space.*

Proof. Let (f_n) be a Cauchy sequence in l^p. Then the sequence $(\tilde{f}_n)$ is a Cauchy sequence in $L^p[0, \infty)$. Therefore $(\tilde{f}_n)$ converges to a function g in $L^p[0, \infty)$ since $L^p[0, \infty)$ is complete. But Proposition 1.8 says that for $1 \leq p < \infty$, $(\tilde{f}_n)$ has a subsequence which converges pointwise almost everywhere to g. [For $p = \infty$, $(\tilde{f}_n)$ converges uniformly to g almost everywhere.] Therefore, it is obvious that g is constant almost everywhere on each $[n - 1, n)$. Let α_n be the value of g on $[n - 1, n)$. Then $f = (\alpha_n) \in l^p$, and it is evident that $f_n \to f$ in the l^p norm. $\qquad\square$

We list the following proposition and inequalities without proofs:

4.3. Proposition. *For $1 \leq p < \infty$, l^p is separable. However, l^∞ is not separable.*

4.4. Hölder Inequality. *Let $p > 1$ and $q > 1$ such that $1/p + 1/q = 1$. If $(x_n) \in l^p$ and $(y_n) \in l^q$, then $(x_n y_n) \in l^1$ and we have*

$$\sum_{n=1}^{\infty} |x_n y_n| \leq \left[\sum_{n=1}^{\infty} |x_n|^p \right]^{1/p} \left[\sum_{n=1}^{\infty} |y_n|^q \right]^{1/q}.$$

4.5. Minkowski Inequality. *Let $p \geq 1$. If (x_n) and (y_n) are in l^p, then*

$$\left[\sum_{n=1}^{\infty} |x_n + y_n|^p \right]^{1/p} \leq \left[\sum_{n=1}^{\infty} |x_n|^p \right]^{1/p} + \left[\sum_{n=1}^{\infty} |y_n|^p \right]^{1/p}.$$

EXERCISES 4

A. Show that $l^p \subset l^q$ if $1 \le p < q \le \infty$.

B. Show that l^p is separable if $1 \le p < \infty$.

C. Show that l^∞ is not separable. *Hint:* If $\{f_1, f_2, \ldots\}$ is a countable set in l^∞ with

$$f_n = (\xi_{n1}, \xi_{n2}, \ldots),$$

define $f \in l^\infty$ by an argument similar to the diagonalization proof of the uncountability of $\mathbb{R}$ (see Proposition 4.3, Chapter Zero).

§5. Hilbert Spaces

One of the most important function spaces in functional analysis is *Hilbert space*, named after the German mathematician David Hilbert (1862–1943), who introduced this space in his work in integral equations, which was published in Göttingen Nachrichten between 1904 and 1910 in six papers. A collection of these papers was published in 1912 under the title *Grundzüge einer allgemeinen Theorie der linearen Integralgleichungen*. These papers are among the most influential published in modern times. Hilbet was concerned with solutions of the integral equation

$$f(x) = \varphi(x) + \int_a^b K(x, t)\varphi(t)\, dt$$

for the unknown function φ. The sequence space l^2 we introduced in the preceding section is a byproduct of these investigations. An axiomatic definition of Hilbert space was given by J. von Neumann (1903–1957) in his "Allgemeine Eigenwertheorie Hermitescher Funktionaloperation" (1929). The axiomatic structure of von Neumann is a realization of the L^2 and l^2 spaces. We shall give the definition of a Hilbert space which is more general than that given by von Nemann. The Hilbert space in the sense of von Neumann will be a separable infinite-dimensional Hilbert space. For the sake of simplicity we introduce here a real Hilbert space.

5.1. Definition. By a *Hilbert space* we mean a Banach space H in which there is defined a function $(x|y)$ on $H \times H$ to $\mathbb{R}$ with the following properties:

  (i) $(x|y) = (y|x)$;
  (ii) $(\alpha x + \beta y|z) = \alpha(x|z) + \beta(y|z)$ for $\alpha, \beta \in \mathbb{R}$;
  (iii) $(x|x) \ge 0$; and
  (iv) $(x|x) = \|x\|^2$.

We call $(x|y)$ the *inner product* of x and y.

5.2. Example. Three examples are immediate:

(a) The Euclidean n-space $\mathbb{R}^n$ is a Hilbert space with respect to the inner product

$$(x|y) = \sum_{j=1}^{n} x_j y_j,$$

where $x = (x_1, \ldots, x_n)$ and $y = (y_1, \ldots, y_n)$.

(b) The space $L^2(E)$ is a Hilbert space with respect to the inner product

$$(f|g) = \int_E f(x)g(x)\, dx.$$

(c) The space l^2 is a Hilbert space with respect to the inner product

$$(x|y) = \sum_{n=1}^{\infty} x_n y_n,$$

where $x = (x_1, x_2, \ldots)$ and $y = (y_1, y_2, \ldots)$.

5.3. Proposition. *If H is a Hilbert space, then for any x, y in H,*

$$|(x|y)| \leq \|x\|\, \|y\|.$$

(Cauchy–Bunyakovsky–Schwarz inequality).

Proof. The equality is clearly valid when $y = 0$. If $y \neq 0$, put $\lambda = \|x\|/\|y\|$. Then

$$0 \leq \|x - \lambda y\|^2 = (x - \lambda y | x - \lambda y)$$

$$= (x|x) - 2\lambda(x|y) + \lambda^2(y|y)$$

$$= 2\|x\|^2 - 2\|x\|\frac{(x|y)}{\|y\|}.$$

Therefore,

$$(x|y) \leq \|x\|\, \|y\|.$$

From this inequality it follows that

$$|(x|y)| \leq \|x\|\, \|y\|. \qquad \square$$

The above inequality applied to the space $L^2(E)$ is the Hölder–Riesz inequality for $p = q = 2$.

Since Hilbert space is a special normed space, it is reasonable to expect that the norm on Hilbert space will have properties which are not common to a norm. One such property is given below.

5.4. Parallelogram Identity. *For all x and y in a Hilbert space,*

$$\|x + y\|^2 + \|x - y\|^2 = 2(\|x\|^2 + \|y\|^2).$$

Proof.

$$\|x + y\|^2 + \|x - y\|^2 = (x + y|x + y) + (x - y|x - y),$$
$$= 2(x|x) + 2(y|y),$$
$$= 2(\|x\|^2 + \|y\|^2). \qquad \square$$

A Hilbert space can be described as a Banach space whose norm satisfies the parallelogram identity. In fact, the inner product is defined by

$$4(x|y) = \|x + y\|^2 - \|x - y\|^2.$$

This is a result obtained by J. von Neumann and P. Jordan in 1935. We leave the proof to the reader as an exercise.

The parallelogram identity can be compared with Euclid's parallel postulate, which states that *through a point not on a line, there is no more than one line parallel to the line.* In this sense, Hilbert space is natural generalization of Euclidean geometry. The following proposition shows that among L^p and l^p spaces, L^2 and l^2 are the only spaces analogous to the Euclidean plane.

5.5. Proposition. $L^p(E)$ $[m(E) > 0]$ *and* l^p *are Hilbert spaces if and only if* $p = 2$.

Proof. It suffices to show that the parallelogram identity is not valid for L^p and l^p if $p \neq 2$. For l^p, we see this easily by applying the parallelogram identity to the vectors

$$e_1 = (1, 0, 0, \ldots),$$
$$e_2 = (0, 1, 0, \ldots).$$

For $L^p(E)$, we take f and g to be two functions such that $\|f\|_p = \|g\|_p = 1$ and $\int_E f(x)g(x)\,dx = 0$. More explicitly, let E_1 and E_2 be two disjoint measurable subsets of E such that

$$m(E_1) = m(E_2) = \frac{m(E)}{2}.$$

(How?) Let $\lambda = m(E_1)$. Set

$$f = \lambda \chi_{E_1},$$
$$g = \lambda \chi_{E_2}.$$

Then f and g do not satisfy the parallelogram identity. $\qquad \square$

5.6. Definition. Let x and y be two elements of a Hilbert space H. Then x and y are said to be *orthogonal* to each other if $(x|y) = 0$; we write $x \perp y$. A subset A of H is called an *orthogonal* set if any two distinct elements in A are orthogonal. An *orthonormal* set is an orthogonal set A with the additional property that $\|x\| = 1$ for every x in A.

5.7. Examples. (a) The standard basis of $\mathbb{R}^n$, $e_1, e_2, \ldots, e_n$, where

$$e_1 = (1, 0, 0, \ldots, 0),$$

$$e_2 = (0, 1, 0, \ldots, 0),$$

$$\ldots\ldots\ldots\ldots\ldots$$

$$e_n = (0, 0, \ldots, 0, 1),$$

is an orthonormal set.

(b) In l^2, $e_1, e_2, \ldots$, where

$$e_1 = (1, 0, 0, \ldots),$$

$$e_2 = (0, 1, 0, \ldots),$$

$$e_3 = (0, 0, 1, \ldots),$$

$$\ldots\ldots\ldots\ldots\ldots$$

is an orthonormal set.

(c) In $L^2[-\pi, \pi]$,

$$\frac{1}{\sqrt{2\pi}}, \quad \frac{\cos x}{\sqrt{\pi}}, \quad \frac{\sin x}{\sqrt{\pi}}, \quad \frac{\cos 2x}{\sqrt{\pi}}, \quad \frac{\sin 2x}{\sqrt{\pi}}, \ldots$$

are orthonormal.

(d) In $L^2[-1, 1]$, the *Legendre polynomials*

$$P_n(x) = \frac{1}{2^n n!} \frac{d^n}{dx^n}(x^2 - 1)^n, \qquad n = 1, 2, \ldots,$$

are orthogonal. Then the following polynomials become orthonormal:

$$\varphi_n(x) = \sqrt{n + \tfrac{1}{2}}\, P_n(x), \qquad n = 1, 2, \ldots.$$

Many more examples of orthonormal sets are known [see Szegö (1959)].

As in Euclidean geometry, we have the following theorem:

5.8. Pythagorean Theorem. Let $x_1, \ldots, x_n$ be an orthogonal set in H and let $x = x_1 + \cdots + x_n$. Then

$$\|x\|^2 = \|x_1\|^2 + \cdots + \|x_n\|^2.$$

Proof.

$$\|x\|^2 = (x_1 + \cdots + x_n | x_1 + \cdots + x_n)$$

$$= \|x_1\|^2 + \cdots + \|x_n\|^2. \qquad \square$$

5.9. Corollary. *If* $\{u_1, \ldots, u_n\}$ *is an orthonormal set, and if* $x = \sum_{k=1}^{n} \alpha_k u_k$, *then*

$$\|x\|^2 = \sum_{k=1}^{n} |\alpha_k|^2,$$

$$\alpha_k = (x|u_k), \qquad 1 \le k \le n.$$

We recall the following concepts from linear algebra:

5.10. Definition. Let V be a vector space, and $x_1, \ldots, x_n \in V$. The set $x_1, \ldots,$ x_n is called *linearly independent* if

$$\alpha_1 x_1 + \cdots + \alpha_n x_n = 0$$

implies $\alpha_1 = \cdots = \alpha_n = 0$. A set $S \subset V$ is *linearly independent* if every finite subset of S is linearly independent.

5.11. Proposition. *Let* $A \subset H$ *be orthonormal. Then* A *is linearly independent.*

Proof. The assertion follows from Corollary 5.9.   □

An orthonormal basis is of great importance in studying finite dimensional spaces $\mathbb{R}^n$. In this connection an orthonormal basis is a set of orthogonal unit vectors whose linear combinations span the whole space. In the infinite-dimensional case the analogue of such a basis is a complete orthonormal set which will be defined below.

5.12. Definition. An orthonormal set A is said to be *complete* if there does not exist a nonzero element in H orthogonal to each element of A. In other words, if $x \in H$, $(x|\varphi) = 0$ for all $\varphi \in A$, then $x = 0$.

A complete orthonormal set is obviously a maximal one. Conversely, if A is a maximal orthonormal set, then A must be complete. For, if $(x|\varphi) = 0$ for all $\varphi \in A$ and $x \ne 0$, then $x/\|x\|$ can be added to A.

A complete orthonormal set is known to exist in any Hilbert space by Zorn's lemma [see Halmos (1960)]. If H is a separable Hilbert space, it is possible to replace this transfinite argument by a constructive method; one such method is the *Gram–Schmidt orthogonalization process* (see §7).

The simplest complete orthonormal set in l^2 is the sequence (e_n) of vectors

$$e_1 = (1, 0, 0, \ldots),$$

$$e_2 = (0, 1, 0, \ldots),$$

$$e_3 = (0, 0, 1, \ldots),$$

$$\ldots\ldots\ldots\ldots\ldots$$

The orthogonality and completeness of this sequence are obvious.

EXERCISES 5

A. Show that if x is orthogonal to $y_1, \ldots, y_n$, then x is orthogonal to any linear combination

$$\alpha_1 y_1 + \cdots + \alpha_n y_n.$$

B. Show that the inner product is continuous; that is, if $x_n \to x$ and $y_n \to y$, then $(x_n | y_n) \to (x | y)$.

C. On $C[-1, 1]$, define the inner product by

$$(f | g) = \int_{-1}^{1} f(x) g(x) \, dx.$$

Show that $C[-1, 1]$ is not a Hilbert space. *Hint:* Show that $C[-1, 1]$ is not complete with respect to the norm $\|f\| = (f|f)^{1/2}$.

D. Let H be a Banach space whose norm satisfies the parallelogram identity. Define

$$(x | y) = \frac{\|x + y\|^2 - \|x - y\|^2}{4}.$$

Show that $(x | y)$ is an inner product which makes H a Hilbert space. *Hint:* To show that

$$(x + y | z) = (x | z) + (y | z)$$

apply the parallelogram identity to parallelograms constructed on the elements: (1) $x, y + z$; (2) $x, y - z$; (3) $z, x + y$; and (4) $z, x - y$. To show that

$$(\alpha x | y) = \alpha (x | y)$$

show first for integers α, then for fractions, and then pass to the limit.

E. If $x_1, x_2, \ldots,$ are orthogonal to an element y and $x = \lim x_n$, then show that x is also orthogonal to y. *Hint:* Exercise B.

F. Apply Zorn's lemma to show that every Hilbert space has a complete orthonormal set.

§6. The Riesz–Fischer Theorem

Henceforth, we shall deal only with a separable Hilbert space, that is, a Hilbert space with a countable dense set.

6.1. Proposition. *If H is a separable Hilbert space, then each orthonormal set in H must be countable.*

Proof. Let D be a countable dense set in H and let A be an orthonormal set in H. Then any two elements φ, ψ in A are at distance $\sqrt{2}$; that is,

$$\|\varphi - \psi\| = \sqrt{2}.$$

For each $\varphi \in A$, there is, since D is dense in H, an element $x_\varphi \in D$ such that

$$\|x_\varphi - \varphi\| < \frac{1}{\sqrt{2}}.$$

Clearly, $x_\varphi \neq x_\psi$ if $\varphi \neq \psi$. Since D is countable, A is at most countable.   □

Thus each orthonormal set in a separable Hilbert space may be expressed as a sequence (φ_n), which may be finite or infinite.

Let (φ_n) be an orthonormal sequence in H and suppose $x \in H$. Let us try to approximate x in the norm as closely as possible by a linear combination

$$c_1 \varphi_1 + \cdots + c_N \varphi_N$$

of the first N elements of the sequence by suitably choosing the coefficients $c_1, \ldots, c_N$, that is, let us find a method of computing the minimum value of

$$\left\| x - \sum_{k=1}^N c_k \varphi_k \right\|$$

for suitable real numbers $c_1, \ldots, c_N$. We have

$$\left\| x - \sum_{k=1}^N c_k \varphi_k \right\|^2 = \left(x - \sum_{j=1}^N c_j \varphi_j \middle| x - \sum_{k=1}^N c_k \varphi_k \right)$$

$$= (x|x) - \sum_{j=1}^N c_j (\varphi_j|x) - \sum_{k=1}^N c_k (x|\varphi_k) + \sum_{k=1}^N \left[\sum_{j=1}^N c_j c_k (\varphi_j|\varphi_k) \right]$$

$$= \|x\|^2 - 2 \sum_{k=1}^N c_k (x|\varphi_k) + \sum_{k=1}^N c_k^2$$

$$= \|x\|^2 - \sum_{k=1}^N (x|\varphi_k)^2 + \sum_{k=1}^N [(x|\varphi_k) - c_k]^2.$$

It is clear that the minimum will be attained when the last term is equal to zero, that is, when

$$c_k = (x|\varphi_k)$$

for $k = 1, \ldots, N$. Thus we obtain the following assertion:

6.2. Proposition. *Let (φ_n) be an orthonormal sequence and let $x \in H$. Then*

$$\left\| x - \sum_{k=1}^N (x|\varphi_k)\varphi_k \right\|^2 = \|x\|^2 - \sum_{k=1}^N (x|\varphi_k)^2$$

(Bessel's identity);

$$\sum_{k=1}^N (x|\varphi_k)^2 \leq \|x\|^2$$

(Bessel's inequality).

6.3. Corollary. *Let (φ_n) be an orthonormal sequence and let $x \in H$. Then*

$$\sum_{k=1}^{\infty} (x|\varphi_k)^2 \leq \|x\|^2.$$

Proof. Since Bessel's inequality holds for any N, we get the inequality in this corollary by letting $N \to \infty$. $\qquad\qquad\square$

We shall also refer to the inequality in the preceding corollary as *Bessel's inequality.*

By analogy with the situation in n-dimensional space $\mathbb{R}^n$, we can expect that in Bessel's inequality the equality is valid whenever the orthogonal sequence in question is complete. We are going to formulate and prove this fact in the following fundamental theorem:

6.4. Theorem. *Let (φ_n) be a complete orthonormal sequence in a Hilbert space H. Then every element x in H admits a series expansion*

$$x = \sum_{n=1}^{\infty} (x|\varphi_n)\varphi_n. \tag{1}$$

Furthermore,

$$\|x\|^2 = \sum_{n=1}^{\infty} (x|\varphi_n)^2. \tag{2}$$

The series expansion (1) means that

$$\left\| x - \sum_{n=1}^{N} (x|\varphi_n)\varphi_n \right\| \to 0$$

as $N \to \infty$. Equation (2), which represents an infinite-dimensional generalization of the Pythagorean theorem, is generally referred to as *Parseval's formula.*

Proof. For the sake of simplicity, let

$$c_n = (x|\varphi_n)$$

for arbitrary $x \in H$. Then by Bessel's inequality (Corollary 6.3) we have

$$\sum_{n=1}^{\infty} c_n^2 \leq \|x\|^2.$$

Then the sequence

$$f_k = \sum_{n=1}^{k} c_n \varphi_n$$

is a Cauchy sequence in H, since for $q > p$

$$f_q - f_p = \sum_{p+1}^{q} c_n \varphi_n$$

and we have

$$\|f_q - f_p\|^2 = \sum_{p+1}^{q} c_n^2$$

which tends to zero as $p \to \infty$ because $\sum_{n=1}^{\infty} c_n^2$ converges. Since H is complete, there is an element $f \in H$ such that

$$f = \lim f_p,$$

that is,

$$f = \sum_{n=1}^{\infty} c_n \varphi_n.$$

We shall show that $f = x$. To do this we observe that for fixed k and $p > k$,

$$(f|\varphi_k) = \lim_{p \to \infty} (f_p|\varphi_k) = \lim_{p \to \infty} \sum_{n=1}^{p} c_n(\varphi_n|\varphi_k)$$

$$= \lim_{p \to \infty} c_k(\varphi_k|\varphi_k) = \lim_{p \to \infty} c_k = c_k.$$

(The first equality is a consequence of Exercise 5B.) It follows that for any k,

$$(x - f|\varphi_k) = 0.$$

Since the orthonormal sequence (φ_n) is complete, we have $x = f$ and

$$x = \sum_{n=1}^{\infty} c_n \varphi_n.$$

Moreover, applying Bessel's identity in Proposition 6.2, we obtain

$$\|x\|^2 - \sum_{n=1}^{p} c_n^2 = \left\| x - \sum_{n=1}^{p} c_n \varphi_n \right\|^2 \to 0$$

as $p \to \infty$; that is,

$$\|x\|^2 = \sum_{n=1}^{\infty} c_n^2. \qquad \square$$

6.5. Definition. The expansion (1) in Theorem 6.4 is called the *generalized Fourier series of* x, and its coefficients

$$c_n = (x|\varphi_n)$$

are called the *generalized Fourier coefficients of* x with respect to the orthonormal sequence (φ_n).

The reader should notice that the generalized Fourier series of x is defined with respect to any orthonormal sequence whether it is complete or not. In the presence of a complete orthonormal sequence the generalized Fourier series of x converges to x with respect to the norm.

The following more general form of Parseval's formula is also valid:

6.6. Proposition. Let (φ_n) be a complete orthonormal sequence in H. Then for $x, y \in H$ we have

$$(x|y) = \sum_{n=1}^{\infty} (x|\varphi_n)(y|\varphi_n).$$

Proof. This follows from (2), Theorem 6.4, by virtue of the relation between the inner product and the norm:

$$4(x|y) = \|x + y\|^2 - \|x - y\|^2. \qquad \square$$

We have given several different theorems the name "Riesz–Fischer theorem." The following theorem is the one originally found by Riesz for the Hilbert space L^2 which motivated the others. For a proof, examine the proof of Theorem 6.4.

6.7. Riesz–Fischer Theorem. Let (φ_n) be an orthonormal sequence in H, and let $(c_n) \in l^2$. Then there corresponds an element $x \in H$ such that

$$x = \sum_{n=1}^{\infty} c_n \varphi_n,$$

where

$$c_n = (x|\varphi_n).$$

If the orthonormal sequence is complete, then the correspondence is unique and

$$\|x\|^2 = \sum_{n=1}^{\infty} c_n^2.$$

If the sequence (φ_n) is orthonormal but not necessarily complete, then instead of Parseval's formula we have only Bessel's inequality.

Although we have not shown that every separable Hilbert space has a complete orthonormal sequence, we shall use this fact in the following discussion. The existence of such a sequence will be proved in the next section.

In a separable Hilbert space, every complete orthonormal sequence has the same number of elements. This can be seen easily. Suppose that the sequence has a finite number n of elements. Then these n elements form a basis for the vector space H by Theorem 6.4. Hence H is an n-dimensional vector space, and hence all complete orthonormal sequences have the same number n of elements. We now give the following definition.

6.8. Definition. The *dimension* of a Hilbert space is the cardinal number of a complete orthonormal set.

Thus if H has a countably infinite complete orthonormal sequence, we say H is of dimension $\aleph_0$ and write dim $H = \aleph_0$.

The Riesz–Fischer theorem immediately implies the following important proposition:

6.9. Proposition. *If H is of dimension $\aleph_0$, then H is isomorphic to the space l^2.*

Two Hilbert spaces H and H' are said to be *isomorphic* if there is a one–one correspondence between their elements such that $x \leftrightarrow x'$ and $y \leftrightarrow y'$ imply that:

(i) $x + y \leftrightarrow x' + y'$;
(ii) $\alpha x \leftrightarrow \alpha x$; and
(iii) $(x|y) = (x'|y')$.

Obviously, two isomorphic Hilbert spaces are isometric if we consider them merely as metric spaces.

Proof of Proposition. Choose an arbitrary complete orthonormal sequence (φ_n) which has an infinite number of elements. Assign to each $x \in H$ the sequence (c_n) of its generalized Fourier coefficients with respect to (φ_n). Then

$$\|x\|^2 = \sum_{n=1}^{\infty} c_n^2 < \infty$$

and hence $(c_n) \in l^2$. This correspondence has the following properties: if

$$x \leftrightarrow (c_n) \quad \text{and} \quad y \leftrightarrow (d_n)$$

then:

(i) $x + y \leftrightarrow (c_n + d_n)$;
(ii) $\alpha x \leftrightarrow (\alpha c_n)$; and
(iii) $(x|y) = \sum_{n=1}^{\infty} c_n d_n$.

Property (iii) is the generalized Parseval formula (see Proposition 6.6). Hence the above correspondence between H and l^2 is an isomorphism. This proves the proposition. $\square$

In the preceding proposition we proved that a separable infinite-dimensional Hilbert space is isomorphic to l^2. Therefore all such Hilbert spaces are essentially different realizations of the same space. This theory of Hilbert space is one of the most important achievements in modern mathematics and is indispensable in theoretical physics, in particular, quantum mechanics. The isomorphism between L^2 and l^2 spaces established above is closely related to the theory of quantum mechanics. Originally, quantum mechanics consisted of two theories. One was Heisenberg's matrix mechanics, and the other was Schrödinger's wave mechanics. However, these two theories are equivalent, which was shown by Schrödinger later. The difference between the two theories reduced to the fact, from the mathematical point of view, that the former used the space l^2 and the latter used the space L^2.

It would lead us beyond the limits of this book to give more details about the general theory of Hilbert space, which can be found in most texts in functional analysis. We close this section with a characterization of bounded linear functions on a Hilbert space. We need the following definition:

6.10. Definition. Let E be a normed space (over $\mathbb{R}$). A function $L: E \to \mathbb{R}$ is said to be a *linear functional* if

$$L(\alpha x + \beta y) = \alpha L(x) + \beta L(y)$$

for all $x, y \in E$ and $\alpha, \beta \in \mathbb{R}$. A linear functional L is said to be *bounded* if there exists a constant $M > 0$ such that

$$|L(x)| \le M \|x\|$$

for all $x \in E$.

Let H be a Hilbert space. The problem is to characterize all bounded linear functionals on H.

Let a be a fixed element in H. The map $L: x \to (x|a)$ is obviously a linear functional. It is bounded since, by the Cauchy–Bunyakovksy–Schwarz inequality,

$$|(x|a)| \le \|a\| \|x\|.$$

We shall now prove the converse of this result. Although the following theorem is true for any Hilbert space, we present here a proof for a separable Hilbert space, in the spirit of this section.

This theorem was discovered independently by M. Fréchet (1907) and F. Riesz (1907a).

6.11. Theorem (Fréchet–Riesz). *Let L be a bounded linear functional on a Hilbert space H. Then there exists a unique element $a \in H$ such that*

$$L(x) = (x|a)$$

for all $x \in H$.

Proof. Let (φ_n) be a complete orthonormal sequence in H. We let $b_n = L(\varphi_n)$. Since L is a linear functional, we have

$$L\left(\sum_{n=1}^{m} b_n \varphi_n \right) = \sum_{n=1}^{m} b_n L(\varphi_n) = \sum_{n=1}^{m} b_n^2.$$

Since L is bounded, there exists a constant $M > 0$ such that

$$L\left(\sum_{n=1}^{m} b_n \varphi_n \right) \le M \left\| \sum_{n=1}^{m} b_n \varphi_n \right\| = M \left(\sum_{n=1}^{m} b_n^2 \right)^{1/2}.$$

Therefore,

$$\sum_{n=1}^{m} b_n^2 \le M \left(\sum_{n=1}^{m} b_n^2 \right)^{1/2}.$$

and hence,

$$\left(\sum_{n=1}^{m} b_n^2\right)^{1/2} \le M.$$

Since this is true for all m, we have $\sum_{n=1}^{\infty} b_n^2 \le M^2 < \infty$. By the Riesz–Fischer Theorem 6.7, there exists an element $a \in H$ such that

$$b_n = (a|\varphi_n)$$

for all $n = 1, 2, \dots$. Now let x be an arbitrary element of H. Since

$$\sum_{n=1}^{m} (x|\varphi_n)\varphi_n \to x$$

as $m \to \infty$ and L is bounded, we have

$$L(x) = \lim_{m\to\infty} L\left(\sum_{n=1}^{m} (x|\varphi_n)\varphi_n\right)$$

by Exercise 6D. Then

$$L(x) = \lim_{m\to\infty} \sum_{n=1}^{m} (x|\varphi_n)L(\varphi_n) = \sum_{n=1}^{\infty} (x|\varphi_n)b_n$$

$$= \sum_{n=1}^{\infty} (x|\varphi_n)(a|\varphi_n) = (x|a)$$

by Proposition 6.6. This completes the proof. ☐

EXERCISES 6

A. Let (φ_n) be an orthonormal sequence in a Hilbert space H. Show that (φ_n) is complete if and only if every element $x \in H$ satisfies Parseval's formula.

B. Interpret the Riesz–Fischer Theorem 6.7 for $L^2[-\pi, \pi]$ and the orthonormal sequence

$$\frac{1}{\sqrt{2}}, \frac{\cos x}{\sqrt{\pi}}, \frac{\sin x}{\sqrt{\pi}}, \dots$$

C. Show that in a separable Hilbert space every complete orthonormal sequence has the same number of elements.

D. Let L be a linear functional on a normed space E. Show that the following statements are equivalent:
   (1) L is bounded;
   (2) L is continuous at the origin 0 of E in the sense that if $x_n \to 0$, then $L(x_n) \to 0$; and
   (3) L is continuous at each point x of E in the sense that if $x_n \to x$, then $L(x_n) \to L(x)$.

For the remaining exercises we need the following notations. Let H be a Hilbert space and $x \in H$. Let $x^\perp$ denote the set of all $y \in H$ which are orthogonal to x. If M is a subset of H, let $M^\perp$ be the set of all $y \in H$ which are orthogonal to every $x \in M$.

E. Show that $x^\perp$ is a closed set and $M^\perp$ is a closed set.

F. Show that $M \subset M^{\perp\perp}$, $M^\perp = M^{\perp\perp\perp}$.

G. If M is a closed vector subspace of H, then $H = M \oplus M^\perp$. That is, every $x \in H$ may be expressed uniquely as the sum $x = x_1 + x_2$ of an element x_1 of M and an element x_2 of $M^\perp$. Furthermore, $M = M^{\perp\perp}$.

H. Prove Theorem 6.11 for a general Hilbert space. *Hint*: Let $M = \{x \in H: L(x) = 0\}$. Show that $M^\perp$ is a vector space of dimension 1 unless $M = H$. Then choose $z \in M^\perp$, $z \neq 0$, and let $y = \alpha z$, where $\alpha = L(z)/(z|z)$.

§7. Orthonormalization

We have already assumed the existence of a complete orthonormal sequence in a separable Hilbert space in the preceding section. The purpose of this section is to demonstrate that such a sequence can be found in any separable Hilbert space.

In order to obtain an orthonormal sequence, a systematic method of orthogonalizing a given nonorthogonal sequence is often used. We introduce here such a method known as the *Gram–Schmidt orthonormalization process*.

Let V be a vector space and $A \subset V$. The set of all linear combinations of elements of A is a vector subspace of V. We denote this subspace by $\langle A \rangle$.

7.1. Gram–Schmidt Orthonormalization Process. *If (x_n) is a sequence of linearly independent vectors in a Hilbert space, then there exists an orthonormal sequence (y_n) such that*

$$\langle x_1, \ldots, x_n \rangle = \langle y_1, \ldots, y_n \rangle$$

for all n.

Proof. The sequence (y_n) will be obtained by induction. Let

$$y_1 = \frac{x_1}{\|x_1\|}.$$

Assume inductively that orthonormal vectors

$$y_1, \ldots, y_{n-1}$$

are already found in such a way that

$$\langle x_1, \ldots, x_k \rangle = \langle y_1, \ldots, y_k \rangle$$

for all $k = 1, \ldots, n - 1$. To construct the next vector y_n, let

$$z = x_n - \sum_{k=1}^{n-1} (x_n | y_k) y_k.$$

Then $z \neq 0$. For ortherwise x_n is a linear combination of $y_1, \ldots, y_{n-1}$ and hence is a linear combination of $x_1, \ldots, x_{n-1}$. Furthermore, z is orthogonal to each $y_1, \ldots, y_{n-1}$. Let $y_n = z/\|z\|$. Then $\{y_1, \ldots, y_n\}$ is orthonormal, and

$$\langle x_1, \ldots, x_n \rangle \subset \langle y_1, \ldots, y_n \rangle.$$

It is easy to show that

$$\langle y_1, \ldots, y_n \rangle \subset \langle x_1, \ldots, x_n \rangle.$$

This completes the proof.   □

We need the following criterion for the completeness of an orthonormal sequence in a Hilbert space:

7.2. Proposition. *Let* (φ_n) *be an orthonormal sequence in a Hilbert space H. Then* (φ_n) *is complete if and only if the vector subspace* $\langle \varphi_1, \varphi_2, \ldots \rangle$ *formed by the linear combinations of the vectors* $\varphi_1, \varphi_2, \ldots$, *is dense in H.*

Proof. Suppose that (φ_n) is complete and $x \in H$. Then by the Riesz–Fischer Theorem 6.7, x is the limit of the sequence of partial sums of the generalized Fourier series of x with respect to (φ_n). Therefore the subspace $\langle \varphi_1, \varphi_2, \ldots \rangle$ is dense in H.

Conversely, suppose that $\langle \varphi_1, \varphi_2, \ldots \rangle$ is dense in H. Suppose that there is an x in H such that $(x|\varphi_n) = 0$ for all $n = 1, 2, \ldots$. Then x is orthogonal to any linear combination $\alpha_1 \varphi_1 + \cdots + \alpha_n \varphi_n$ of these vectors, that is, if $y \in \langle \varphi_1, \varphi_2, \ldots \rangle$, then $(x|y) = 0$. Since $\langle \varphi_1, \varphi_2, \ldots \rangle$ is dense in H and $x \in H$, there is a sequence (y_n) in $\langle \varphi_1, \varphi_2, \ldots \rangle$ converging to x. Therefore,

$$(x|x) = \lim(x|y_n) = 0.$$

This implies that $x = 0$, and hence (φ_n) is complete.   □

7.3. Proposition. *Let H be a Hilbert space. Then H is separable if and only if H admits a complete orthonormal sequence.*

Proof. Let $\{x_1, x_2, \ldots\}$ be a countable dense set in H. It is an easy exercise to show that there is a subsequence (y_n), which is linearly independent, and

$$\langle x_1, x_2, \ldots \rangle = \langle y_1, y_2, \ldots \rangle.$$

[In fact, we need only eliminate from the sequence (x_n) all elements x_k which are linear combinations of elements x_j with smaller indices, $j < k$.] Applying the Gram–Schmidt process to the subsequence (y_n), we get an orthonormal sequence (φ_n) such that

$$\langle y_1, y_2, \ldots \rangle = \langle \varphi_1, \varphi_2, \ldots \rangle.$$

Since $\{x_1, x_2, \ldots\}$ is dense in H, $\langle y_1, y_2, \ldots \rangle$ is dense in H, and hence $\langle \varphi_1, \varphi_2, \ldots \rangle$ is dense in H. This proves that (φ_n) is a complete orthonormal sequence in H.

Conversely, suppose (φ_n) is a complete orthonormal sequence in H. By Proposition 7.2, $\langle \varphi_1, \varphi_2, \ldots \rangle$ is dense in H. We observe that the countable set of all combinations of $\varphi_1, \varphi_2, \ldots$ with rational coefficients is dense in $\langle \varphi_1, \varphi_2, \ldots \rangle$; thus H is separable.   □

EXERCISE 7

A. Let V be the vector subspace of $L^2[0, 1]$ of polynomials of degree at most 3. Apply the Gram–Schmidt process to the basis $\{1, x, x^2, x^3\}$.

§8. Completeness of the Trigonometric System

In this section we shall prove the orthonormal sequence

$$\frac{1}{\sqrt{2\pi}}, \quad \frac{\cos x}{\sqrt{\pi}}, \quad \frac{\sin x}{\sqrt{\pi}}, \quad \frac{\cos 2x}{\sqrt{\pi}}, \quad \frac{\sin 2x}{\sqrt{\pi}}, \ldots, \tag{*}$$

is complete in the Hilbert space $L^2[-\pi, \pi]$.

A *trigonometric polynomial* is a finite sum of the form

$$T(x) = a_0 + \sum_{n=1}^{N} (a_n \cos nx + b_n \sin nx).$$

We recall that the Fourier series of f is a trigonometric series

$$f \sim \frac{a_0}{2} + \sum_{n=1}^{\infty} (a_n \cos nx + b_n \sin nx),$$

where

$$a_n = \frac{1}{\pi} \int_{-\pi}^{\pi} f(x) \cos nx \, dx, \qquad n = 0, 1, 2, \ldots,$$

$$b_n = \frac{1}{\pi} \int_{-\pi}^{\pi} f(x) \sin nx \, dx, \qquad n = 0, 1, 2, \ldots,$$

(see §2, Chapter I).

8.1. Lemma. *For any $\delta > 0$ and $\eta > 0$ there is a trigonometric polynomial $T(x)$ such that:*

(i) $T(x) \geq 0$;
(ii) $\int_{-\pi}^{\pi} T(x) \, dx = 1$; *and*
(iii) $T(x) \leq \eta$, $\delta \leq |x| \leq \pi$.

Proof. Take

$$T_n(x) = \frac{(1 + \cos x)^n}{\int_{-\pi}^{\pi} (1 + \cos x)^n \, dx} = \frac{(\cos(x/2))^{2n}}{\int_{-\pi}^{\pi} (\cos(x/2))^{2n} \, dx}.$$

Then conditions (i) and (ii) are obviously satisfied. Furthermore,

$$T_n(x) < \frac{(\cos(\delta/2))^{2n}}{\int_0^{\delta/2} (\cos(x/2))^{2n}\, dx} < \frac{2}{\delta}\left(\frac{\cos(\delta/2)}{\cos(\delta/4)}\right)^{2n} \to 0$$

if $\delta \le x \le \pi$ and $n \to \infty$. Therefore, T_n also satisfies condition (iii) for sufficiently large n. $\square$

8.2. Lemma. *Let $f \in C[-\pi, \pi]$ be such that the Fourier coefficients of f are all 0. Then f is identically zero.*

Proof. Extend the domain of f to the whole real line $\mathbb{R}$ by periodicity with period 2π by $f(x) = f(x + 2\pi)$ if $x \in (-\pi, \pi]$. For each $y \in \mathbb{R}$ and each trigonometric polynomial $T(x)$, we have

$$\int_{-\pi}^{\pi} f(x + y)T(x)\, dx = \int_{y-\pi}^{y+\pi} f(x)T(x - y)\, dx = 0.$$

Suppose that there is a point ξ in $(-\pi, \pi)$ such that $f(\xi) = c \ne 0$. We may assume $c > 0$. Since f is continuous on $[-\pi, \pi]$, there is a $\delta > 0$ such that

$$f(x) > \frac{c}{2}$$

throughout $(\xi - \delta, \xi + \delta) \subset (-\pi, \pi)$. Let $M = \sup\{|f(x)|: x \in \mathbb{R}\}$. Take $T(x)$ as in Lemma 8.1. Then for $x \in (\xi - \delta, \xi + \delta)$ we have

$$0 = \int_{-\pi}^{\pi} f(x + \xi)T(x)\, dx$$

$$\ge \frac{c}{2}\int_{-\pi}^{\pi} T(x)\, dx - M\left(\int_{-\pi}^{-\delta} + \int_{\delta}^{\pi}\right)T(x)\, dx$$

$$= \frac{c}{2} - \left(\frac{c}{2} + M\right)\left(\int_{-\pi}^{-\delta} + \int_{\delta}^{\pi}\right)T(x)\, dx > \frac{c}{2} - \left(\frac{c}{2} + M\right)2\pi\eta.$$

This is a contradiction if η is sufficiently small. Therefore f must be identically zero. $\square$

8.3. Proposition. *Let $f \in L^1[-\pi, \pi]$ be such that the Fourier coefficients of f are all 0. Then f is identically zero.*

Proof. Define $F: [-\pi, \pi] \to \mathbb{R}$ by

$$F(x) = \int_0^x f(t)\, dt.$$

Then F is continuous on $[-\pi, \pi]$. Let $A_0, A_1, A_2, \ldots, B_1, B_2, \ldots,$ be the

Fourier coefficients of F. Then for $n \geq 1$

$$A_n = \frac{1}{\pi} \int_{-\pi}^{\pi} F(x) \cos nx \, dx$$

$$= -\frac{1}{n\pi} \int_{-\pi}^{\pi} f(x) \sin nx \, dx \quad \text{(integration by parts)}$$

$$= -\frac{b_n}{n} = 0.$$

Similarly, $B_n = 0$ for $n \geq 1$. Then the Fourier coefficients of $F - A_0/2$ are all 0. Since $F - A_0/2$ is continuous on $[-\pi, \pi]$, by Lemma 8.2, $F - A_0/2$ is identically zero; that is, $F \equiv A_0/2$. Therefore f is identically zero.   □

8.4. Corollary. *The trigonometric system*

$$\frac{1}{\sqrt{2\pi}}, \quad \frac{\cos x}{\sqrt{\pi}}, \quad \frac{\sin x}{\sqrt{\pi}}, \quad \frac{\cos 2x}{\sqrt{\pi}}, \quad \frac{\sin 2x}{\sqrt{\pi}}, \dots,$$

is a complete orthonormal sequence in $L^2[-\pi, \pi]$.

We now restate the substance of the Riesz–Fischer Theorem 6.7 in the form which it takes for the Fourier series of L^2 functions.

8.5. Riesz–Fischer Theorem. *Let $f \in L^2[-\pi, \pi]$. Then the Fourier series of f converges to f in the L^2 norm; that is,*

$$\int_{-\pi}^{\pi} \left| f(x) - \left[\frac{a_0}{2} + \sum_{n=1}^{N} (a_n \cos nx + b_n \sin nx) \right] \right|^2 dx \to 0$$

as $n \to \infty$.

We also state Parseval's formula for the trigonometric system.

8.6. Parseval's Formulas. *Let $f \in L^2[-\pi, \pi]$. Then*

$$\frac{1}{\pi} \int_{-\pi}^{\pi} f^2(x) \, dx = \tfrac{1}{2}a_0^2 + \sum_{n=1}^{\infty} (a_n^2 + b_n^2).$$

More generally, let $f, F \in L^2[-\pi, \pi]$. Then

$$\frac{1}{\pi} \int_{-\pi}^{\pi} f(x) F(x) \, dx = \tfrac{1}{2}a_0 A_0 + \sum_{n=1}^{\infty} (a_n A_n + b_n B_n),$$

where the A's and B's are the Fourier coefficients of F.

Proof. The first formula is a restatement of Parseval's formula in Theorem 6.4. For the second formula, let

$$s_k = \frac{a_0}{2} + \sum_{n=1}^{k} (a_n \cos nx + b_n \sin nx),$$

$$S_k = \frac{A_0}{2} + \sum_{n=1}^{k} (A_n \cos nx + B_n \sin nx).$$

Then

$$\frac{1}{\pi}(f - s_k | F - S_k) = \frac{1}{\pi}[(f|F) - (f|S_k) - (F|s_k) + (s_k|S_k)]$$

$$= \frac{1}{\pi}(f|F) - \left[\tfrac{1}{2}a_0 A_0 + \sum_{n=1}^{k} (a_n A_n + b_n B_n) \right].$$

But

$$|(f - s_k | F - S_k)| \le \|f - s_k\|_2 \|F - S_k\|_2 \to 0$$

as $k \to \infty$. Therefore,

$$\frac{1}{\pi}(f|F) = \tfrac{1}{2}a_0 A_0 + \sum_{n=1}^{\infty} (a_n A_n + b_n B_n). \qquad \square$$

EXERCISE 8

A. The Riemann–Lebesgue Theorem. *Let f be integrable on $[a, b]$. Then*

$$\lim \int_a^b f(x) \cos nx \, dx = \lim \int_a^b f(x) \sin nx \, dx = 0.$$

Hint: First consider a step function. Then approximate the integrable function f by a step function with respect to the L^1 norm.

§9. Isoperimetric Problem

An interesting application of the Parseval formula to the isoperimetric problem is as follows:

> Show that among all simple plane curves with a given arc length the largest area is enclosed by the circle.

It is easy to conjecture that the curve should be a circle. Many proofs have been given of this result since J. Steiner's geometric argument in 1839. Steiner proved that no curve different from the circle can be a solution to the problem. His proof left unclear whether the circle has this extremum property.

We present here a proof given by A. Hurwitz in 1902.

9.1. Isoperimetric Theorem. *Among all simple closed plane curves with a given arc length, the circle encloses the largest area. In other words, if L is the arc length of a simple closed curve C and A is the area bounded by C, then*

$$L^2 - 4\pi A \geq 0 \qquad \text{(isoperimetric inequality)}.$$

The equality holds only when C is a circle.

Proof. We consider a parametric representation of a simple closed curve C with arc length L. For convenience, we choose our parameter s as arc length:

$$C: \begin{cases} x = x(s), \\ y = y(s). \end{cases}$$

Then

$$0 \leq s \leq L; \qquad x(0) = x(L); \qquad y(0) = y(L).$$

It is easy to show that $x(s)$ and $y(s)$ satisfy the Lipschitz condition.

In fact, we can see from Figure 6.7 that the arc length is greater than the corresponding cord length over $[s_1, s_2]$; that is,

$$|x(s_2) - x(s_1)| \leq |s_2 - s_1| \qquad \text{and} \qquad |y(s_2) - y(s_1)| \leq |s_2 - s_1|.$$

This shows that both $x(s)$ and $y(s)$ are differentiable, and the derivatives $x'(s)$ and $y'(s)$ are bounded (see Proposition 5.2, Chapter V). Moreover, they are absolutely continuous. Hence,

$$[x'(s)]^2 + [y'(s)]^2$$

is integrable, and the arc length can be computed by

$$l = \int_0^l \sqrt{[x'(s)]^2 + [y'(s)]^2} \, ds,$$

where l is arc length of the curve on $[0, l] \subset [0, L]$. Hence

$$[x'(s)]^2 + [y'(s)]^2 = 1.$$

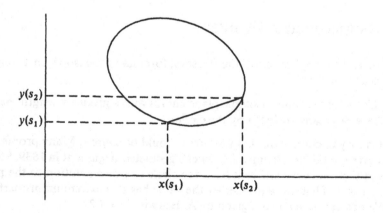

Figure 6.7

In order to utilize the Parseval formula, we have to find Fourier series for $x(s)$ and $y(s)$, but their domain is $[0, L]$ rather than $[0, 2\pi]$. For this reason, we change the parameter s by introducing a new parameter $t = 2\pi s/L$. Then the curve C becomes

$$C: \begin{cases} x = f(t), \\ y = g(t), \end{cases}$$

with

$$0 \le t \le 2\pi; \qquad f(0) = f(2\pi); \qquad g(0) = g(2\pi).$$

The reader should be warned that

$$\int_0^{2\pi} \sqrt{[f'(t)]^2 + [g'(t)]^2}\, dt$$

is not the arc length L of the curve C (why?).

First, we note that

$$[f'(t)]^2 + [g'(t)]^2 = \left(\frac{L}{2\pi}\right)^2 ([x'(s)]^2 + [y'(s)]^2) = \left(\frac{L}{2\pi}\right)^2.$$

Hence

$$\int_0^{2\pi} ([f'(t)]^2 + [g'(t)]^2)\, dt = \frac{L^2}{2\pi}.$$

On the other hand, the enclosed area is

$$A = \int_0^{2\pi} f(t)g'(t)\, dt$$

by a standard calculus formula from Green's theorem.

To obtain isoperimetric inequality, we express A in terms of the Fourier coefficients of $f(t)$ and $g(t)$. Write

$$f(t) \sim \frac{a_0}{2} + \sum_{k=1}^{\infty} (a_k \cos kt + b_k \sin kt),$$

$$g(t) \sim \frac{A_0}{2} + \sum_{k=1}^{\infty} (A_k \cos kt + B_k \sin kt).$$

Then the Fourier series of $f'(t)$ and $g'(t)$ are

$$f'(t) \sim \sum_{k=1}^{\infty} k(b_k \cos kt - a_k \sin kt),$$

$$g'(t) \sim \sum_{k=1}^{\infty} k(B_k \cos kt - A_k \sin kt).$$

By Parseval's formulas, we obtain the area

$$A = \int_0^{2\pi} f(t)g'(t)\, dt = \pi \sum_{k=1}^{\infty} k(a_k B_k - b_k A_k)$$

and

$$\frac{L^2}{2\pi} = \int_0^{2\pi} ([f'(t)]^2 + [g'(t)]^2)\, dt = \pi \sum_{k=1}^{\infty} k^2(a_k^2 + b_k^2 + A_k^2 + B_k^2).$$

Hence

$$L^2 - 4\pi A = 2\pi^2 \sum_{k=1}^{\infty} [k^2(a_k^2 + b_k^2 + A_k^2 + B_k^2 - 2k(a_k B_k - b_k A_k)]$$

$$= 2\pi^2 \sum_{k=1}^{\infty} [(ka_k - B_k)^2 + (kb_k + A_k)^2 + (k^2 - 1)(A_k^2 + B_k^2)].$$

This establishes that

$$L^2 - 4\pi A \geq 0.$$

If C is a circle, it is obvious that

$$L^2 - 4\pi A = 0.$$

Conversely, if $L^2 - 4\pi A = 0$, then

$$ka_k - B_k = 0; \qquad kB_k + A_k = 0; \qquad (k^2 - 1)(A_k^2 + B_k^2) = 0;$$

and hence

$$a_k = b_k = A_k = B_k = 0 \qquad \text{for} \quad k > 1; \qquad a_1 = B_1; \quad b_1 = -A_1.$$

In this case, the Fourier coefficients of the continuous functions

$$f(t) - \left(\frac{a_0}{2} + a_1 \cos t + b_1 \sin t\right),$$

$$g(t) - \left(\frac{A_0}{2} - B_1 \cos t + A_1 \sin t\right),$$

are all 0. It follows from Lemma 8.2 that

$$f(t) = \frac{a_0}{2} + a_1 \cos t + b_1 \sin t,$$

$$g(t) = \frac{A_0}{2} - B_1 \cos t + A_1 \sin t.$$

Thus

$$\left(x - \frac{a_0}{2}\right)^2 + \left(y - \frac{A_0}{2}\right)^2 = a_1^2 + b_1^2.$$

It's a circle!

§10. Remarks on Fourier Series

The question of pointwise convergence of the Fourier series of a given function f was not discussed in this book. One reason is that our interest in the topic has been based on L^2 approximation rather than on the pointwise convergence of the Fourier series. Another reason for avoiding this question of convergence is that it is extremely difficult to answer in a limited time and in a limited space. We recommend the following books for a further study of Fourier series:

G.H. Hardy and W.W. Rogosinski, *Fourier Series*, 3rd edn. (1956);
A. Zygmund, *Trigonometric Series* (1959); and
R.E. Edwards, *Fourier Series* (1967).

The difficulties of pointwise convergence of Fourier series can be pointed up by the fact that *there is a Lebesgue integrable function whose Fourier series diverges everywhere*. Such an example was constructed by the Russian mathematician A.N. Kolmogoroff [see Zygmund (1959), vol. I, pp. 310–314].

The Dirichlet theorem, §2, Chapter I, can be replaced by the following theorem. [See, for example, Hardy and Rogosinski (1956), p. 42.]

10.1. Jordan's Test. *Let $f \in L^1$ be of bounded variation on some neighborhood of a point x. Then the Fourier series of f converges to $\frac{1}{2}[f(x^+) + f(x^-)]$.*

The Riesz–Fischer Theorem 8.5 states that if $f \in L^2$, then its Fourier series converges to f in the L^2 norm. Related to this result, N.N. Lusin (1915) posed the following problem:

Does the Fourier series of any function in L^2 converge almost everywhere?

Over a period of fifty years, an enormous amount of effort was expended on this problem. In 1966, L. Carleson answered Lusin's question affirmatively in the paper, "On convergence and growth of partial sums of Fourier series."

10.2. The Carleson Theorem. *Let $f \in L^2[-\pi, \pi]$. Then the Fourier series of f converges almost everywhere to f.*

A remarkable feature of Carleson's work is that it used no new techniques. The Carleson theorem has been generalized by R.A. Hunt (1970) to the following form:

10.3. The Carleson–Hunt Theorem. *Let $f \in L^p[-\pi, \pi]$, $1 < p \leq \infty$. Then the Fourier series of f converges almost everywhere to f.*

We refer the interested reader to the following monograph: C.J. Mozzochi, *On the Pointwise Convergence of Fourier Series* (1971).

APPENDIX
The Development of the Notion of the Integral

by Henri Lebesgue*

Gentlemen:

Foregoing technical developments, we are going to examine as a whole the successive modifications and enrichments of the notion of the integral and the appearance of other concepts used in recent research concerning functions of real variables.

Before Cauchy, there was no definition of the integral in the actual sense of the word "definition." One was limited to saying which areas it was necessary to add or subtract to obtain the integral.

For Cauchy a definition was necessary, because with him appeared the concern for rigor which is characteristic of modern mathematics. Cauchy defined continuous functions and the integrals of these functions in nearly the same way as we do now. To arrive at the integral of $f(x)$, it sufficed for him to form the sums (see Figure A.1)

$$S = \sum f(\xi_i)(x_{i+1} - x_i), \tag{1}$$

which surveyors and mathematicians have used for centuries for approximating areas, and to deduce from this the integral $\int_a^b f(x)\, dx$ by passage to the limit.

Although this passage was obviously legitimate to those starting with a notion of area, Cauchy had to prove that the sum S actually tended toward a limit under the conditions which he considered. An analogous necessity is imposed each time that one replaces an experimental notion with a purely logical definition. It should be added that the interest of the defined object is

* From a conference in Copenhagen May 8, 1926, at la Société Mathématique. We wish to acknowledge Dr. Paul Ricoeur, Editor, *Revue de Métaphysique et de Morale* for allowing us to translate Lebesgue (1926) for this book.

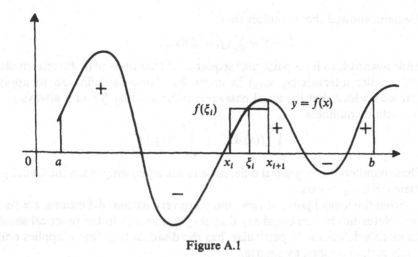

Figure A.1

no longer evident; it can only result from the study of the properties of this object.

This is the price of logical progress. What Cauchy did is considerable enough to have a philosophical meaning. It is often said that Descartes reduced geometry to algebra. I would say more readily that, by employing coordinates, he reduced all geometries to that of the straight line and that this geometry, in giving us the notions of continuity and irrational number, has permitted algebra to attain its actual scope.

In order that the reduction of all geometries to the geometry of the straight line be achieved, it was necessary to eliminate a certain number of notions related to geometries of higher dimensions such as length of a curve, area of a surface, and volume of a body. Precisely here lies the progress which Cauchy realized. After him, it sufficed that arithmeticians construct the linear continuum with the aid of natural numbers to accomplish the arithmetization of the science.

And now, should we limit ourselves to doing analysis? No. Indeed, all that we will do can be translated into arithmetical language, but if one were to refuse to have direct, geometric, intuitive insights, if one were reduced to pure logic which does not permit a choice among everything that is exact, one would hardly think of many questions, and certain notions, the majority of those notions which we are going to examine today, for example, would escape us completely.

For a long time, certain discontinuous functions were integrated; Cauchy's definition still applied to these integrals, but it was natural to investigate, as Riemann did, the exact scope of this definition.

If f_i and $\bar{f}_i$ designate the lower and upper bounds of $f(x)$ on (x_i, x_{i+1}), then S lies between

$$\underline{S} = \sum \underline{f}_i(x_{i+1} - x_i) \quad \text{and} \quad \bar{S} = \sum \bar{f}_i(x_{i+1} - x_i).$$

Riemann showed that it suffices that

$$\bar{S} - \underline{S} = \sum (\bar{f}_i - \underline{f}_i)(x_{i+1} - x_i)$$

tends toward zero for a particular sequence of partitions of (a, b) into smaller and smaller intervals (x_i, x_{i+1}) in order for Cauchy's definition to apply. Darboux added that the usual passages to the limit by $\underline{S}$ and $\bar{S}$ always give two definite numbers

$$\underline{\int_a^b} f(x)\, dx, \qquad \overline{\int_a^b} f(x)\, dx.$$

These numbers are in general different and are equal only when the Cauchy–Riemann integral exists.

From the logical point of view, these are very natural definitions, are they not? Nevertheless, one could say that they are useless in the practical sense. Riemann's definition, in particular, has the disadvantage that it applies only rarely and, in a sense, by chance.

Indeed, it is evident that the partitioning of (a, b) into smaller and smaller intervals (x_i, x_{i+1}) makes the differences $\bar{f}_i - \underline{f}_i$ smaller and smaller if $f(x)$ is continuous, and by virtue of this continuous process it is clear that this partitioning causes $\bar{S} - \underline{S}$ to tend toward zero if there are only a few points of discontinuity. However, there is no reason to hope that the case will be the same for a function discontinuous everywhere. So, in effect, taking smaller and smaller intervals (x_i, x_{i+1}), that is to say, values of $f(x)$ related to values of x which come closer and closer together, in no way guarantees that one takes values of $f(x)$ whose differences become less and less.

Let us proceed according to the goal to be attained: to gather or group values of $f(x)$ which differ by little. It is clear, then, that we must partition not (a, b), but rather the interval $(\underline{f}, \bar{f})$ bounded by the lower and upper bounds of $f(x)$ on (a, b). We do this with the aid of numbers y_i differing amont themselves by less than ε; we are led, for example, to consider values of $f(x)$ defined by

$$y_i \leq f(x) \leq y_{i+1}.$$

The corresponding values of x form a set E_i; in the case of Figure A.2, this set E_i is made up of four intervals. With a certain continuous function $f(x)$, it might be formed by an infinite number of intervals. With an arbitrary function, it might be very complicated. But, no matter—it is this set E_i which plays the role analogous to that of the interval (x_i, x_{i+1}) in the definition of the integral of continuous functions, since it makes known to us the values of x which give to $f(x)$ values differing very little.

If η_i is any number chosen between y_i and y_{i+1},

$$y_i \leq \eta_i \leq y_{i+1},$$

the values of $f(x)$ for the points of E_i differ from η_i by less than ε. The number η_i will play the role which was assumed by $f(\xi_i)$ in (1); as for the role of the length or measure $x_{i+1} - x_i$ of the interval (x_i, x_{i+1}), this will be played by a

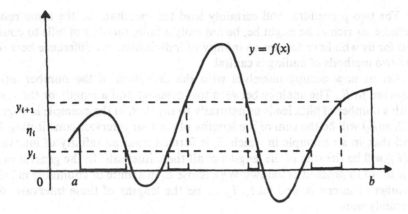

Figure A.2

measure $m(E_i)$, which we will assign to the set E_i in a moment. We form in this manner the sum

$$S = \sum \eta_i m(E_i). \tag{2}$$

But first let us look at what we have just done and, in order to understand it better, repeat it in other terms.

The geometers of the seventeenth century considered the integral of $f(x)$—the word "integral" had not yet been invented, but that is hardly important—as the sum of an infinity of indivisibles,[1] each of which is an ordinate, positive or negative, of $f(x)$. Very well! We have simply grouped the indivisibles of comparable size; we have, as one says in algebra, made the collection or reduction of similar terms. It may again be said that, with Riemann's procedure, one attempted to sum the indivisibles by taking them in the order in which they were furnished by the variation of x. One operated as did a merchant without a method who counted coins and bills randomly in the order in which they fell into his hand, while we operate like the methodical merchant who says:

I have $m(E_1)$ pennies worth $1 \cdot m(E_1)$;
I have $m(E_2)$ nickels worth $5 \cdot m(E_2)$; and
I have $m(E_3)$ dimes worth $10 \cdot m(E_3)$;

etc., and thus I have altogether

$$S = 1 \cdot m(E_1) + 5 \cdot m(E_2) + 10 \cdot m(E_3) + \cdots.$$

[1] In the context of areas, *indivisibles* are "infinitely narrow" rectangles of "infinitesimal" area. Leibniz used the symbol dx to denote the "width" of an indivisible, so that the "area" of an indivisible of length y was given by the product $y\, dx$. He then introduced the symbol $\int y\, dx$ for the "sum" or "integral" of the areas of the indivisibles which gives the area of a given region—S.B.C.

The two procedures will certainly lead the merchant to the same result because, as rich as he might be, he has only a finite number of bills to count; but for us who have to sum an infinity of indivisibles, the difference between the two methods of adding is capital.

Let us now occupy ourselves with the definition of the number $m(E_i)$ attached to E_i. The analogy between this measure and a length, or the same with a number of bills, leads us naturally to say that, in the example in Figure A.2, $m(E_i)$ will be the sum of the lengths of the four intervals constituting E_i, and that, in an example in which E_i is formed from an infinity of intervals, $m(E_i)$ will be the sum of the lengths of all these intervals. In the general case, it leads us to proceed as follows: We enclose E_i in a finite or countably infinite number of intervals, and let $l_1, l_2, \ldots$ be the lengths of these intervals. We certainly want

$$m(E_i) \leq l_1 + l_2 + \cdots.$$

If we look for the greatest lower bound of the second member for all possible systems of intervals which can serve to cover E_i, this bound will be an upper bound for $m(E_i)$. For this reason we denote it by $\overline{m(E_i)}$ and we have

$$m(E_i) \leq \overline{m(E_i)}. \tag{3}$$

If C is the set of points of (a, b) not included in E_i, we have similarly

$$m(C) \leq \overline{m(C)}.$$

Now, we obviously wish to have

$$m(E_i) + m(C) = m((a, b)) = b - a.$$

Therefore we must have

$$m(E_i) \geq b - a - \overline{m(C)}. \tag{4}$$

The inequalities (3) and (4) give then the upper and lower bounds of $m(E_i)$. One can easily see that these two inequalities are never contradictory. When the lower and upper bounds of E_i are equal, $m(E_i)$ is defined, and we say then that E_i is measurable.[2]

A function $f(x)$ for which the sets E_i are measurable for all y_i is called measurable. For such a function, formula (2) defines a sum S. One can easily prove that, when one varies the choice of y_i in such a way that ε tends toward zero, the sum S approaches a definite limit which is by definition $\int_a^b f(x)\, dx$.[3]

---

[2] The method of defining the measure of sets used here is that of C. Jordan (*Cours d'Analyse de l'École Polytechnique*, vol. 1) but with this modification essential to our aim: that we enclose the set E_i to be measured in intervals which may be infinite in number, whereas C. Jordan always used only a finite number of intervals. This use of a countable infinity in place of integers is suggested by the endeavors of Borel who, moreover, himself used this idea in particular for a definition of measure (*Leçons sur la Théorie des Fonctions*).

[3] *Comptes Rendus Acad. Sci. Paris*, **129**, 1909. Definitions equivalent to that of the text were proposed by various authors. The most interesting are due to W.H. Young (*Philos. Trans. Roy. Soc. London*, **204**, 1905; *Proc. London Math. Soc.*, 1910). See also, for example, the notes by Borel and by F. Riesz (*Comptes Rendus Acad. Sci. Paris*, **154**, 1912).

This first extension of the notion of integral led to many others. Let us suppose that it is a question of integrating a function $f(x, y)$ in two variables. We proceed exactly as before. We assign to it sets E_i which are now points in the plane and no longer the points on a line. To these sets we must now attribute a plane measure; this measure is deduced from the area of the rectangles

$$\alpha \leq x \leq \beta, \qquad \gamma \leq y \leq \delta,$$

entirely in the same manner as linear measure is deduced from the length of intervals. With the measure defined, formula (2) will give the sum S, from which the integral is deduced by passage to the limit.

The definition which we have considered thus extends itself immediately to functions of several variables. Here is another extension which applied equally well whatever the number of variables, but which I state only in the case where it is a question of integrating $f(x)$ on (a, b).

I have said that it is a matter of forming the sum of indivisibles represented by the various ordinates of the points x, $y = f(x)$. A moment ago, we grouped these indivisibles according to their size. Let us now restrict ourselves to grouping them according to their sign; we will have to consider the plane set E_p of those points, the ordinates of which are positive, and the set E_n of points with negative ordinates. For the simple case in which $f(x)$ is continuous, before Cauchy, as I recalled in the beginning, one wrote

$$\int_a^b f(x)\, dx = \text{area}(E_p) - \text{area}(E_n).$$

This leads us to write

$$\int_a^b f(x)\, dx = m_s(E_p) - m_s(E_n),$$

m_s designating a plane measure. This new definition is equivalent to the preceding one. It brings us back to the intuitive method before Cauchy, but the definition of measure has given it a solid logical foundation.

We thus know two ways of defining the integral of a function of one or more variables, and that we know without having to consider the more or less complicated form of the domain of integration, because the domain D intervenes only as follows: The sets E_i of our first definition and the sets E_p, E_n of the second are formed by taking values of the function f only on the points of D.

Since the choice of the domain of integration D enters only in the formation of the sets E_i, or E_p and E_n, it is clear that we could just as well agree to form these sets E_i, E_n, E_p by taking into consideration only the values assumed by f on the points of a given set E, and we will have hence defined the integral of f extended to the set E.

In order to make precise the scope of this new extension of the notion of integral, let us recall that our definitions require that f be measurable, that is to say, that the sets E_i be measurable for the first definition, and that E_p and

E_n be measurable for the second, and, in view of this, E must also be measurable. We thus know how to define the integral extended to a measurable set of a measurable and bounded function on this set. I have, in effect, implicitly supposed thus far that we are dealing with bounded functions.

What would have to be changed in the first manner of definition if the function f to be integrated were not bounded? The interval $(\underline{f}, \bar{f})$ would no longer be finite; an infinity of numbers y_i would be needed to divide it into intervals of length at most equal to ε, so there would be an infinity of sets E_i and the sum S of formula (2) would now be a series. In order not to be stopped at the outset, we must assume that the series S is convergent for the first choice of the numbers y_i that we would make; but, if S exists for one choice of y_i, it exists for all choices of y_i, and the definition of the integral applies without modification.

The name of summable functions has been given to all functions which can be integrated by the indicated procedures, that is to say, to all measurable functions for which the sums S have a meaning. Every bounded measurable function is summable; and as no one has up to now succeeded in naming a nonmeasurable function, one could say that, up to now, practically every bounded function has an integral.[4] On the contrary, there exist very simple unbounded functions which are not summable. Thus, one must not be astonished that our notion of integral still reveals itself insufficient in certain questions.

We have just extended the notion of integral to unbounded functions by starting with the first of our definitions; the second leads to the same result. But for this it is necessary to enlarge the notion of measure in such a way that it applies not only to bounded sets, which we thus far considered solely, but also to sets of points extending to infinity. I mention this second method of proceeding only because it is also related to another extension of the definite integral in which the interval, the domain, the set on which the integral is extended, is no longer presumed finite, as we have done up to now, but may go to infinity.

I limit myself to just an indication, because I will not be considering in what follows this extension of the integral concept. It is for the same reason that I am content with mentioning briefly the research, still very original, undertaken by a young man killed in the war, R. Gateaux, who intended to define the operation of integration for functions of infinitely many variables. This research, which was continued by Paul Lévy and by Norbert Wiener, is not without relation to the axiomatic studies undertaken by M. Fréchet and by P.J. Daniell with the aim of extending the notion of integral to abstract sets.[5] Fréchet and Daniell proposed furthermore to apply to abstract sets not

---

[4] See §4, Chapter III—S.B.C.

[5] R. Gateaux, *Bull. Soc. Math. France*, 1919; P. Lévy, *Leçons d'Analyse Fonctionelle*, 1922; N. Wiener, *Proc. London Math. Soc.*, 1922; M. Fréchet, *Bull. Soc. Math. France*, 1915: P.J. Daniell, *Ann. of Math.*, 1918 and 1919.

only the definitions of which I have spoken thus far, but also a further extension of the definite integral, to which we shall be led soon by the notion of indefinite integral, which we are now going to examine.

One ordinarily calls the indefinite integral of a function $f(x)$ the function $F(x)$ defined by

$$F(x) = C + \int_a^x f(x)\, dx. \tag{5}$$

We do not adhere to this name but give rather to the words "indefinite integral" their original meaning. Originally, the two names "definite integral" and "indefinite integral" applied to the same expression $\int_a^b f(x)\, dx$. But the integral was called "definite" when it was a question of a given, determined, or defined interval (a, b); and the integral was "indefinite" when (a, b) was variable, undetermined, undefined, or, if one wishes, indefinite.

It is, in short, by a veritable abuse of language that one calls $F(x)$ the indefinite integral of $f(x)$. If we remark in addition that when one studies $F(x)$ it is always to obtain properties of $\int_a^b f(x)\, dx$, that it is actually $\int_a^b f(x)\, dx$ which one studies through $F(x)$, one will be led to say: I call the indefinite integral of $f(x)$ the function

$$\varphi(a, b) = \int_a^b f(x)\, dx = F(b) - F(a). \tag{5'}$$

There are between an indefinite integral and the corresponding definite integral the same relations and same differences as between a function and a particular value taken on by this function. Furthermore, if we represent by D the interval (a, b) of integration, we may say that the indefinite integral is a function, the argument of which is the domain D,

$$\psi(D) = \varphi(a, b).$$

From these reflections it clearly results that, relative to a function of two variables $f(x, y)$, one must not take for the indefinite integral, as is sometimes done, the function

$$F(X, Y) = c_1(x) + c_2(y) + \int_a^x \int_\beta^y f(x, y)\, dx\, dy. \tag{6}$$

If one limits oneself to considering rectangular domains

$$a \le x \le b, \quad c \le y \le d,$$

one must take for the indefinite integral the function of four variables

$$\varphi(a, b; c, d) = F(b, d) + F(a, c) - F(a, d) - F(b, c). \tag{7}$$

But if one wishes to consider all the domains of integration, since the most general domain cannot be determined by a finite number of parameters, however large the number, it becomes necessary to give up ordinary functions to represent the correspondence between a domain D and the integral extended to this domain and to study directly the function

$$\psi(D) = \iint_D f(x, y) \, dx \, dy$$

for which the argument D is a domain. It is this function which we will call the indefinite integral of $f(x, y)$. Or rather, since we have also defined the integral of f extended to a measurable set E, we will consider the indefinite integral as a set function which will have been defined for all measurable sets.[6]

In all that has been said up to now, there are, to be sure, only questions of language or of naming; but these questions would not have been asked if we had not acquired a new concept. It is for this reason that one should not be surprised that the new language has allowed one to give all possible meaning to facts perceived first of all in the case of the function $F(x)$ of formula (5). One has succeeded, in particular, in characterizing set functions which are indefinite integrals byy two properties: complete additivity and absolute continuity.[7]

When a set function possesses these two properties, it is the indefinite integral of a function f which depends on 1, 2, 3, ... variables according to whether the sets E are formed with the aid of the points on a line, in a plane, in ordinary space, etc. In order to have a uniform language and notation, let us say that f is a point function, $f(P)$; we write

$$\psi(E) = \int_E f(P) \, dm(E). \tag{8}$$

The function $f(P)$ is entirely determined by $\psi(E)$ to the extent that one can arbitrarily modify f on the points of an arbitrary set of measure zero without its ceasing to have $\psi(E)$ for an indefinite integral. And one can obtain $f(P)$ starting with $\psi(E)$, except on points of a set of measure zero, by the following procedure.

Let P be the point at which we wish to calculate f; we take for the domain of integration Δ an interval with center P, or a circle with center P, or a sphere with center P—according to whether we are dealing with the case of the line, plane, or space—and we form the ratio $\psi(\Delta)/m(\Delta)$. Then, let Δ tend to zero and we have

$$\lim_{\Delta \to 0} \frac{\psi(\Delta)}{m(\Delta)} = f(P). \tag{9}$$

---

[6] *Ann. Sci. de l'École Normale Supérieure*, 1910.

[7] These terms are due, respectively, to de la Vallée-Poussin (*Intégrales de Lebesgue, Fonctions d'Ensemble, Classes de Baire*, Paris, 1916) and G. Vitali (*R. Acc. Sci. Torino*, 1908). A function of a measurable set is absolutely continuous if, when E varies in such a way that $m(E)$ tends toward zero, $\psi(E)$ also tends toward zero. "Complete additivity" is a synonym for "countable additivity" (see §2, Chapter III)—S.B.C.

This result evidently generalizes the classical theorem according to which, if $f(x)$ is continuous, the function $F(x)$ of formula (5) admits f as its derivative; our procedure of calculating $f(P)$ is indeed, in effect, a sort of differentiation of the set function $\psi(E)$.

This manner of differentiation was considered quite a long time ago. Cauchy[8] calls "coexistent quantities" those quantities determined at the same time, that is, by the same conditions. If, for example, one has a non-homogeneous body, nonhomogeneous in composition and density, and if one considers a domain D of this body, the volume of D, the mass of D, the quantity of heat necessary to elevate by one degree the temperature of D supposed isolated, all are coexistent quantities. These are functions $V(D)$, $M(D)$, $Q(D)$ of the domain.

It is not by happy chance that we arrive here at functions of domains. If one reflects on it, one quickly sees that every magnitude of physics is related not to a point, but to an extended body, that it is a function of a domain, at least insofar that it is a matter of directly measurable magnitudes. The body to be considered will not, however, always be a body of our customary space; it could be a body in a purely mathematically conceived space if, in the determination of the envisaged magnitude, there intervene nonspatial variables such as time, temperature, etc. But this is of little importance; directly measurable magnitudes—mass, quantity of heat, quantity of electricity, for example—are functions of a domain and not functions of a point.

Physics meanwhile also considers magnitudes associated with points, such as speed, tension, density, specific heat; but these are derived magnitudes which one defines accurately most often by the ratio or the limit of the ratio of two coexistent quantities:

$$\text{Density} = \frac{\text{mass}}{\text{volume}}, \quad \text{Specific heat} = \frac{\text{quantity of heat}}{\text{mass}},$$

that is to say, by taking the derivative of a magnitude with respect to a coexistent quantity.

Thus physics, and consequently geometry, leads to the consideration of functions of a domain and their differentiation, just as does analysis of functions of real variables. Similarly the functions of a domain have, in physics, a somewhat more primordial role than point functions. Why then do physicists not speak of these functions? Because mathematicians have not yet studied them and because algebra has notation neither for the domains, nor for the functions of domains. Thus one sees the physicist limit himself to considering special domains depending only on certain parameters, in such a way that the domain function to be considered is reduced to a function of parameters. This is, moreover, exactly what a mathematician does when, instead of considering the definite integral of $f(x, y)$ in all of its generality, he limits himself to considering the functions $F(X, Y)$, $\varphi(a, b; c, d)$ of formulas (6) and (7).

[8] *Exercises d'Analyse et de Physique Mathématique*, vol. 2, Paris, 1840–1847, pp. 188–229.

We remark furthermore that formula (8) establishes a connection between the set functions $\psi(E)$, which are indefinite integrals, and point functions $f(P)$, which are dependent upon algebra. This formula (8) thus furnishes a sort of notation for certain set functions. But when one examines the two conditions required for a function to be an indefinite integral, one cannot doubt that physical quantities are among the class of functions susceptible to this notation.

These reflections on the nature of physical quantities may have allowed you to understand more precisely the interest and the importance of the notions which we have encountered. They show, in particular, that the operation of differentiation which appears in formula (9) is not the only one to be considered, that one can always consider the differentiation of a function $\psi(E)$ with respect to a coexistent function $p(E)$, whether or not it is the measure $m(E)$.

One question now quickly comes to mind: Can one also replace the function $m(E)$ with a given function $p(E)$ in the definition of the integral? In this there is no difficulty. We will first replace formula (2) by

$$S = \sum \mu_i p(E_i),$$

if first the sets E_i belong to the family of those sets for which the function $p(E)$ is defined—that is, the function to be integrated must be measurable with respect to $p(E)$ in order for the series S to be convergent, that is, f must be summable with respect to $p(E)$. This being presumed, the definition of the integral of $f(P)$ with respect to $p(E)$,

$$\int f(P) \, dp(E)$$

is obtained as before if the function $p(E)$ possesses a certain property which one expresses in saying that $p(E)$ must be of bounded variation.[9]

We have just arrived at a new and very considerable extension of the notion of integral in taking the formal point of view of the mathematician;

[9] $p(E)$ is said to be of bounded variation if, in whichever manner one partitions E into a countably infinite number of pairwise disjoint sets $E_1, E_2, \ldots$, the series $\sum |p(E_i)|$ is convergent.

The notion of functions of bounded variation was first introduced by C. Jordan for functions of one variable.

The only set functions $p(E)$ to be considered in these theories are additive functions, that is, those for which one has

$$p(E_1 + E_2 + \cdots) = p(E_1) + p(E_2) + \cdots,$$

$E_1, E_2, \ldots$ being pairwise disjoint. If the additivity is complete, that is, if the sequence $E_1, E_2, \ldots$ can be chosen arbitrarily, $p(E)$ is necessarily of bounded variation. In effect, the order of the sets being unimportant, the series $p(E_1) + p(E_2) + \cdots$ must remain convergent whatever the order; that is, the series $\sum |p(E_i)|$ is convergent.

No attempts have been made up to now to get rid of the condition that $p(E)$ be of bounded variation. One ought to remark besides that if $p(E)$ were not of bounded variation, one could find a continuous function $f(P)$ for which, nevertheless, our definition of integral would not apply.

the point of view of the physicist leads even more naturally to the same result, at least for continuous functions $f(P)$. One could similarly say that the physicists have always considered only integrations with respect to domain functions.

Suppose, for example, that one wishes to calculate the quantity of heat $\varphi(D)$ necessary to elevate by one degree the temperature of a body D of which we spoke above. One must divide D into small partial bodies $D_1, D_2, \ldots$ of masses $M(D_1), M(D_2), \ldots$, choose from each a point $P_1, P_2, \ldots$, and choose for an approximate value of $\varphi(D)$ the sum

$$f(P_1)M(D_1) + f(P_2)M(D_2) + \cdots,$$

$f(P)$ designating the specific heat at P. This is to say that we are calculating $\varphi(D)$ by the formula

$$\varphi(D) = \int_D f(P)\, dM(E).$$

In its general form the new integral was defined only in 1913 by Radon; it was, meanwhile, known since 1894 for the particular case of a continuous function of a single variable. But its first inventor, Stieltjès, was led to it by research in analysis and arithmetic, and he presented it in a purely analytical form which masked its physical significance so much that it required much effort to understand and recognize what is now evident. The history of these efforts cites the names of F. Riesz, H. Lebesgue, W.H. Young, M. Fréchet, C. de la Vallée-Poussin; it shows that we competed in ingenuity and in perspicacity, but also in blindness.[10]

And yet, mathematicians always considered Stieltjès–Radon integrals. The curvilinear integral $\int_C f(x, y)\, dx$ is one of these integrals, relative to a function defined in terms of the length of the projection onto the x axis of arcs of C; the integral $\iint_S f(x, y, z)\, dx\, dy$ involves in the same way a set function defined in terms of areas of S projected onto the xy-plane.

In truth, these integrals most often present themselves in groups

$$\int_C f(x, y)\, dx + g(x, y)\, dy,$$

$$\int_S f(x, y, z)\, dx\, dy + g(x, y, z)\, dx\, dz + h(x, y, z)\, dz\, dx.$$

If one thinks also of integrals considered for the definition of lengths of curves or areas of surfaces,

$$\int_C (dx^2 + dy^2 + dz^2)^{1/2}, \qquad \iint_S [(dx\, dy)^2 + (dy\, dz)^2 + (dz\, dx)^2]^{1/2},$$

[10] J. Radon, *Sitz. Kais. Ak. Wiss. Vienna*, vol. 122, Section IIa, 1913; T.J. Stieltjès, *Ann. Fac. Sci. Toulouse*, 1894; F. Riesz, *Comptes Rendus Acad. Sci. Paris*, 1909; H. Lebesgue, *ibid.*, 1910; W.H. Young, *Proc. London Math. Soc.*, 1913; M. Fréchet, *Nouv. Ann. des Math.*, 1909; de la Vallée-Poussin, *op. cit.*

One will be led to say that it is also convenient to study modes of integration in which there appear several set functions $p_1(E)$, $p_2(E)$, ... This study remains entirely for the future, although Hellinger and Toeplitz have utilized certain summations with respect to several set functions.[11]

We have thus far considered integration, definite or indefinite, as an operation furnishing a number, defined or variable, by a sort of generalized addition. We are placed with the point of view of quadratures. But one may also consider the integration of a continuous function as furnishing a function, just like the most simple of integrations of differential equations. It is this point of view of primitive functions which we will now consider.

Finding the primitive function $F(x)$ of a given function $f(x)$ is finding the function, determined to an additive constant, when it exists, which admits $f(x)$ as its derivative. It is this problem that we are going to study.

But first we remark that the preceding reflections lead to formulating the problem in a much more general fashion: Given a function $f(P)$ which is the derivative with respect to a known function $p(E)$ of an unknown function $\psi(E)$, find the primitive function $\psi(E)$ of $f(P)$.

If, for example, we are dealing with a continuous function $f(x)$ and if $m(E)$ is the measure, the primitive function would no longer be the function $F(x)$ of formula (5), but the indefinite integral $\int_E f(x)\,dx$.

I can only mention this general problem which has not been studied; I am content with remarking that the Stieltjès integral would be very insufficient for resolving it. This integral has, in effect, only been defined for the hypothesis that $p(E)$ is of bounded variation, and one may certainly speak of differentiation with respect to a function $p(E)$ which is not of bounded variation.

The theory of summable functions furnishes the following result related to the case in which $p(E)$ is the measure $m(E)$: When the derivative $f(P)$ is summable, the antiderivative of f is one of its primitive functions. I say one of its primitive functions because one still does not know very well now this general problem of primitive functions must be posed in order for it to be determined.[12]

Let us leave aside these questions, which I speak of only in order to show how much there remains to be done, and let us show how much has been done in the search for the primitive function $F(x)$ of $f(x)$, thanks above all to Arnaud Denjoy.

I have just said that, when $f(x)$ is summable, integration furnishes $F(x)$ by formula (5). Suppose that, on (a, b), $f(x)$ fails to be summable only at a single point c. Then integration gives us $F(x)$ on $(a, c - \varepsilon)$ for arbitrarily small ε and hence on the whole interval (a, c); it also gives $F(x)$ on $(c + \varepsilon, b)$ and hence completely on (c, b). And taking into account the continuity of $F(x)$ at the point c, we have $F(x)$ on the whole interval (a, b). By such considerations

[11] See, for example, *J. Reine Angew. Math.*, **144**, 1914, pp. 212–238.

[12] See on this subject the notes of Fubini and Vitali, appearing 1915–1916, in *Atti Rend. R. Acc. Lincei*.

of continuity,[13] one sees that, if one knows $F(x)$ on every interval which contains no point of a set E in its interior or at its extremeties, one can deduce $F(x)$ by an operation which I shall designate by A on every interval adjacent to E, that is, on every interval having its end points in E but having no points of E in its interior.

Suppose now that one knows $F(x)$ on intervals (α, β) adjacent to a set E, that the sum $\sum [F(\beta) - F(\alpha)]$ is convergent, and that $f(x)$ is summable on E.[14] Then it suffices to say that the primitive function must result from the contribution of E and the intervals adjacent to E in order to be led to the formula

$$F(x) - F(a) = \left\{ \int_E f \, dx + \sum [F(\beta) + F(\alpha)] \right\}_a^x ,$$

the braces of the second member indicating that one must utilize there only points between a and x. From this formula there results the determination of $F(x)$, thanks to an operation which I will designate by B.

The preceding results mark the extreme points which I reached in my thesis, and I must say that I indicated them only somewhat by chance, because I did not at all suspect the importance given to them by Denjoy. Relying on Baire's results, Denjoy shows that, if $f(x)$ is a derivative function on (a, b), then:

(1) The points for which $f(x)$ is not summable form a set E_1 which is not dense in (a, b); an operation O_1 of type A determines $F(x)$ on intervals adjacent to E_1.

(2) Next, there exists a set E_2 formed from points of E_1 and not dense in E_1, on the adjacent intervals of which one can calculate $F(x)$ by an operation O_2 of type B.

(3) Next, there exists a set E_3 formed from points of E_2 and not dense in E_2, on the adjacent intervals of which one can calculate $F(x)$ by an operation O_3 of type $B, \ldots$.

If it turns out that after an infinite sequence of operations $O_1, O_2, \ldots$, one has not yet found $F(x)$ on the entire interval (a, b), the points of (a, b) which are not interior points of intervals on which one has defined $F(x)$ form a set E_ω, and an operation of type A, the operation O_ω, furnishes F on intervals adjacent to E_ω. One considers next, if it is necessary, operations $O_{\omega+1}$, $O_{\omega+2}, \ldots$ of type B, followed by an operation $O_{2\omega}$ of type A, followed by operations of type B, etc.

[13] It is the introduction of these conditions of continuity which very considerably differentiates the problem of primitive functions from that of quadratures.

[14] It is convenient to remark that these hypotheses are not contradictory, the same as if E is assumed to be the set of points on which $f(x)$ is not summable in an interval (a, b) considered. For the determination of points of nonsummability on (a, b) it is necessary, in effect, to take into account all points of (a, b), whether they belong to E or not; whereas summability on E is a condition occurring only on the points of E.

And Denjoy, using now classical arguments of Cantor and Bendixson, proves that this procedure will finally give us $F(x)$ on the entire interval (a, b) after a finite or countably infinite number of operations.

This operative procedure, certainly complicated, but just as natural, in principal, as those previously envisaged, was called by Denjoy "totalization."

Totalization solves entirely the problem of finding the primitive function $F(x)$ of a given function $f(x)$; it permits at the same time the determination of $F(x)$ knowing only a derived number[15] f of $F(x)$ and no longer its derivative. I shall not dwell on these beautiful results; the most important fact for us is that totalization, by a long detour, furnishes us with a new extension of the concept of definite integral. Every time, in effect, that totalization applies to a function $f(x)$ and gives a corresponding function $F(x)$, we can attach to $f(x)$ an integral, thanks to formulas (5) and (5′).[16]

Gentlemen, I end now and thank you for your courteous attention; but a word of conclusion is necessary. This is, if you will, that a generalization made not for the vain pleasure of generalizing, but rather for the solution of problems previously posed, is always a fruitful generalization. The diverse applications which have already taken the concepts which we have just examined prove this superabundantly.

[15] Dini's derivative—S.B.C.

[16] The detailed memoirs of Denjoy appeared from 1915 to 1917 in the *Journal de Math.*, in the *Bull. Soc. Math. France*, and in the *Ann. Sci. de l'École Normale Supérieure*.

Bibliography

Ampère, A.M. (1906). Recherches sur quelques points de la théorie des fonctions derivées qui conduissent à une nouvelle démonstration de la série de Taylor, *Journal de l'École Polytechnique* (Paris), **13**:148–181.

Arzelà, C. (1885). Sulla integrazione per serie, *Rendiconti Reale Accademia dei Lincei* (Rome), **1**:532–537, 566–569.

Banach, S. (1923). Sur le problème de mesure, *Fundamenta Mathematicae*, **4**:7–33.

Banach, S. (1925). Sur les lignes rectifiables et les surfaces dont l'aire est finie, *Fundamenta Mathematicae*, **7**:225–236.

Banach, S. (1932). *Théorie des Opérations Linéaires*, Monografie Matematyczne, Warsaw; reprinted by Chelsea, New York, 1963.

Banach, S., and Tarski, A. (1924). Sur la décomposition des ensembles de points en parties respectivement congruentes, *Fundamenta Mathematicae*, **6**:243–277.

Behrend, F.A. (1960). Crinkly curves and choppy surfaces, *American Mathematical Monthly*, **67**:971–973.

Boas, R.P. (1972). *A Primer of Real Functions*, 2nd edn., The Mathematical Association of America.

Bolzano, B.P.J.N. (1817). *Reine Analytischer Beweis des Lehrsatzes*, Gottlieb Haase, Prague.

Bolzano, B.P.J.N. (1930). *Funktionenlehre, Schriften*, vol. 1, Karl Petr, Prague.

Borel, E. (1895). *Leçons sur la Théorie des Fonctions*, Gauthier-Villars, Paris, 3rd edn. of the same, 1928.

Bunyakovsky, V. (1859). Sur quelques inéqualites concernant les intégrales ordinaires et les intégrales aux différences finies, *Memoires de l'Académie de St. Petersbourg*, **VII**, 1, no. 9.

Cantor, G. (1882). Über unendliche, linear Punktmannichfaltigkeiten, *Mathematische Annalen*, **20**:113–121.

Cantor, G. (1884). De la puissance des ensembles parfaits de points, *Acta Mathematica*, **4**:381–392.

Carleson, L. (1966). On convergence and growth of partial sums of Fourier series, *Acta Mathematica*, **116**:135–157.

Cauchy, A.L. (1821). *Cours d'Analyse de l'École Royal Polytechnique*, Chez Debure Frères, Paris; *Oeuvres Complètes* (*Complete Works*), Académie des Sciences, Ser. 2, 3 (1897).

Cauchy, A.L. (1823). *Résumé des Leçons Données a l'École Royal Polytechnique sur le Calcul Infinitésimale*, Chez Debure Frères, Paris; *Oeuvres Complètes (Complete Works)*, Académie des Sciences, Ser. 2, **4** (1899).

Chae, S.B., and Peck, V. (1973). A generalization of Steinhaus–Kemperman theorem, *Notices, American Mathematical Society*, #709-B24.

Cunningham, F., Jr., and Grosman, N. (1971). On Young's inequality, *American Mathematical Monthly*, **78**:781–783.

Darboux, G. (1875). Mémoire sur la théorie des fonctions discontinues, *Annales Scientifiques de l'École Normale Supérieure*, (2) **4**:57–112.

Dekker, T.J., and de Groot, J. (1956). Decompositions of a sphere, *Fundamenta Mathematicae*, **43**:185–194.

Dini, U. (1878). *Foundamenti per la teorica della funzioni di variabili reali*, Pisa. German translation, *Grundlagen für eine Theorie der Funktionen einer veränderlichen reelen Grösse*, Teubner, Leipzig, 1892.

Dirichlet, P.G. (1829). Sur la convergence des séries trigonométriques qui servent a représenter une fonction arbitraire entre des limites données, *J. Reine Angewandte Mathematik*, **4**:157–169.

Du Bois-Reymond, P. (1875). Versuch einer Classifikation der Willkürlichen Funktionen reeller Argumente, *J. Reine Angewandte Mathematik*, **79**:21–37.

Edwards, R.E. (1967). *Fourier Series, A Modern Introduction*, 2 vols., Holt, Rinehart, and Winston, New York.

Egoroff, D.F. (1911). Sur les suites de fonctions measurables, *Comptes Rendus de l'Académie des Sciences* (Paris), **152**:244–246.

Euclid (circa 300 B.C.). *The Thirteen Books of Euclid's Elements*, vol. 1, Sir Thomas Heath (ed.), Dover, New York, 1956.

Faber, G. (1910). Über stetige Funktionen II, *Mathematische Annalen*, **69**:372–433.

Fatou, P. (1906). Séries trigonométriques et séries de Taylor, *Acta Mathematica*, **30**:335–400.

Fischer, E. (1907a). Sur la convergence en moyenne, *Comptes Rendus de l'Académie des Sciences* (Paris), **144**:1022–1024.

Fischer, E. (1907b). Aplications d'un théorème sur la convergence moyenne, *Comptes Rendus de l'Académie des Sciences* (Paris), **144**:1148–1151.

Fréchet, M. (1907). Sur les ensembles de fo nctions et les opérations linéaires, *Comptes Rendus de l'Académie des Sciences* (Paris) **144**:1414–1416.

Fréchet, M. (1913). Pri le funkcia ekvacio $f(x + y) = f(x) + f(y)$, *Enseignement Mathématique*, **15**:390–393.

Fréchet, M. (1928). *Les Espaces Abstraits*, Gauthier-Villars, Paris.

Fourier, J. (1822). *La Théorie Analytique de la Chaleur*, Didot, Paris. English translation: *The Analytic Theory of Heat*, Cambridge University Press, Cambridge, 1878.

Fubini, G. (1907). Sugli integrali multipli, *Rendiconti Reale Accademia dei Lincei* (Rome), **5**:608–614.

Gillman, L., and Jerison, M. (1960). *Rings of Continuous Functions*, D. van Nostrand, New York; reprinted by Springer-Verlag, New York, 1976.

Grattan-Guinness, I. (1970). *The Development of the Foundations of Mathematical Analysis from Euler to Riemann*, M.I.T. Press, Cambridge, MA.

Halmos, P.R. (1950). *Measure Theory*, van Nostrand, New York.

Halmos, P.R. (1960). *Naive Set Theory*, van Nostrand, New York.

Hardy, G.H., Littlewood, J.E., and Polya, G. (1959). *Inequalities*, Cambridge University Press, Cambridge.

Hardy, G.H., and Rogosinski, W.W. (1956). *Fourier Series*, 3rd edn., Cambridge University Press, Cambridge.

Hausdorff, F. (1914). *Grundzüge der Mengenlehre*, Leipzig; reproduced by Chelsea, New York, 1955.

Hawkins, T. (1970). *Lebesgue's Theory of Integration, Its Origin and Development*, University of Wisconsin Press, Madison, WI.

Hewitt, E. (1960). *Theory of Functions of a Real Variable*, preliminary edn., Holt, Rinehart, and Winston, New York.

Hilbert, D. (1912). *Grundzüge einer allgemeinen Theorie der linearen Integralgleichungen*, Teubner, Leipzig and Berlin.

Hobson, E.W. (1909). On some fundamental properties of Lebesgue integrals in a two-dimensional domain. *Proceedings of the London Mathematical Society* (2), 8:22–39.

Hölder, O. (1998). Über einen Mittelwertsatz, *Göttinger Nachrichten*: 38–47.

Hunt, R.A. (1970). Almost everywhere convergence of Walsh–Fourier series of L^2 functions, *Actes Congrès Intern. Math., Nice*, 2:655–661.

Jordan, C. (1881). Sur la série de Fourier, *Comptes Rendus de l'Académie des Sciences* (Paris), 92:228–230.

Kakutani, S., and Oxtoby, J.C. (1950). A non-separable translation invariant extension of the Lebesgue measure space, *Annals of Mathematics*, 2:580–590.

Kennedy, H.C. (1972). Who discovered Boyer's Law? *American Mathematical Monthly*, 79:66–67.

Kuratowski, K. (1966). *Topology I*, Academic Press, New York.

Lebesgue, H. (1901). Sur une généralisation de l'intégrale définite, *Comptes Rendus de l'Académie des Sciences* (Paris), 132:1025–1028.

Lebesgue, H. (1902). Intégrale, longueur, aire, *Annali di Mathematica Pura ed Applicata* (3), 7:231–359.

Lebesgue, H. (1904). *Leçons sur l'Intégration et la Recherche des Fonctions Primitives*, Gauthier-Villars, Paris.

Lebesgue, H. (1905). Sur les fonctions représentables analytiquement, *Journal of Mathematics*, Ser. 16, 1:139–216.

Lebesgue, H. (1926). Sur le développement de la notion d'intégrale, *Matematisk Tidsskrift*, Copenhagen; reprinted in *Revue de Métaphysique et de Morale*, 34 (1927): 149–167; in Spanish in *Revista Matemática Hispano–Americano*, Ser. 2 (1927); in English in this book and Lebesgue (1966), pp. 178–194.

Lebesgue, H. (1928). *Leçons sur l'Intégration et la Recherche des Fonctions Primitives*, 2nd edn., Gauthier-Villars, Paris.

Lebesgue, H. (1966). *Measure and the Integral*, edited, with Biographical Essay, by K.O. May, Holden-Day, San Francisco.

Levi, B. (1906a). Sopra l'integrazione delle serie, *Rendiconti Reale Instituto Lombardo di Scienze e Lettere* (Milano) (2), 36:775–780.

Levi, B. (1906b). Sur principio di Dirichlet, *Rendiconti del Circolo Matematico di Palermo*, 22:293–359.

Lusin, N.N. (1912). Sur les propriétés des fonctions mesurables, *Comptes Rendus de l'Académie des Sciences* (Paris), 154:1688–1690.

Minkowski, H. (1896). *Geometrie der Zahlen*, Teubner, Leipzig; reprinted by Chelsea, New York, 1953.

Mozzochi, C.J. (1971). *On the Pointwise Convergence of Fourier Series*, Lecture Notes No. 199, Springer-Verlag, New York.

Munroe, M.E. (1953). *Introduction to Measure and Integration*, Addison-Wesley, Cambridge, MA.

Natanson, I.P. (1955). *Theory of Functions of a Real Variable*, vol. 1, Ungar, New York.

Natanson, I.P. (1960). *Theory of Functions of a Real Variable*, vol. 2, Ungar, New York.

Osgood, W.F. (1897). Nonuniform convergence and the integration of series term by term, *American Journal of Mathematics*, 19:155–190.

Oxtoby, J.C. (1971). *Measure and Category*, Springer-Verlag, New York.

Riemann, B. (1866). Über die Darstellbarkeit einer Funktion durch eine trigonometrische Reihe, *Abh. Gesell. Wiss. Göttingen*, 13; *Math. Klasse*, 87–132; *Gesammelte Mathematische Werke*, 2nd edn., and *Nachtrage*, Teubner, Leipzig, 1902, pp. 227–271.

Riesz, F. (1905). Sur un théorème de M. Borel, *Comptes Rendus de l'Académie des Sciences* (Paris), 140:224–226.

Riesz, F. (1907a). Sur les systèmes orthogonaux de fonctions, *Comptes Rendus de l'Académie des Sciences* (Paris), 144:615–619.

Riesz, F. (1907b). Über orthogonale Funktionensysteme, *Nachrichten von der Königl. Gesell. Wiss. Göttingen, Math. Klasse*: 116–122.

Riesz, F. (1909). Sur les suites de fonctions mesurables, *Comptes Rendus de l'Académie des Sciences* (Paris), 148:1303–1305.

Riesz, F. (1910). Untersuchungen über Systeme integrirbarer Funktionen, *Mathematische Annalen*, 69:449–497.

Riesz, F. (1920). Sur l'intégrale de Lebesgue, *Acta Mathematica*, 42:191–205.

Riesz, F. (1932). Sur l'existence de la derivée des fonctions monotones et sur quelques problèmes qui s'y rattachent, *Acta Litt. Sci. Math. Szeged*, 5:208–221.

Riesz, F., and Sz.-Nagy, B. (1956). *Functional Analysis*, English edn., Ungar, New York.

Robinson, R.M. (1947). On the decomposition of sphere, *Fundamenta Mathematicae*, 34:246–260.

Rubel, L.A. (1963). Differentiability of monotone functions, *Colloquium Mathematicum*, 10:276–279.

Saks, S. (1937). *Theory of the Integral*, 2nd revised edn., Monografie Matematyczne, vol. 7, Warsaw; reprinted by Dover, New York, 1964.

Schwarz, H.A. (1885). Über ein die Flächen kleinsten Flächeninhalts betreffendes Problem der Variationsrechnung, *Acta Soc. Scient. Fenn.*, 15:315–362.

Sierpinski, W. (1953). On the congruence of sets and their equivalence by finite decomposition. In *Congruence of Sets and Other Monographs*, Chelsea, New York, 1967.

Solovay, R. (1970). A model of set theory in which every set of reals is Lebesgue measurable, *Annals of Mathematics*, (2) 92:1–56.

Spivak, M. (1967). *Calculus*, Benjamin, Menlo Park, CA.

Steinhaus, H. (1920). Sur les distances des points dans les ensembles de mesure positive, *Fundamenta Mathematicae*, 1:93–104.

Stromberg, K. (1979). The Banach–Tarski paradox, *American Mathematical Monthly*, 86:151–161.

Suslin, M.Ya. (1917). Sur une definition des ensembles mesurables B sans nombre transfinis, *Comptes Rendus de l'Académie des Sciences* (Paris), 164:88–91.

Szegö, G. (1959). *Orthogonal Polynomials*, Colloqium Publications, vol. 23a, American Mathematical Society, Providence, RI.

Tonelli, L. (1909). Sull'integrazione per parti, *Rendiconti Reale Accademia dei Lincei* (Rome), 18:246–253.

Van der Waerden, B.L. (1930). Ein einfaches Beispiel einer nichtdifferenzierbaren stetigen Funktion, *Zeitschrift für Mathematik und Physik*, 32:474–475.

Van Vleck, E.B. (1908). On non-measurable sets of points, with an example, *Transactions of the American Mathematical Society*, 9:237–244.

Vitali, G. (1905). *Sul problema della misura dei gruppi de punti di una retta*, Memorie della Accademia della Scienze dell'Instituto de Bologna.

Vitali, G. (1908). Sui gruppi di punti e sulle funzioni di variabili reali, *Atti della R. Accad. Sci. di Torino*, 43:75–92.

von Neumann, J. (1929). Allgemeine Eigenwerttheorie Hermitescher Funktionaloperation, *Mathematische Annalen*, 102:49–131.

von Neumann, J. (1950). *Functional Operators*, vol. 1, Princeton University Press, Princeton, NJ.

Wilansky, A. (1967). Additive functions. In *Lectures on Calculus*, K.O. May (Ed.), Holden-Day, San Francisco, pp. 97–124.

Wilder, R.L. (1965). *Introduction to the Foundations of Mathematics*, 2nd edn., Wiley, New York.

Young, G.C., and Young, W.H. (1911). On the existence of a differential coefficient, *Proceedings of the London Mathematical Society* (2), 9:325–335.

Young, W.H. (1912). On classes of summable functions and their Fourier series, *Proceedings of the Royal Society* (London) (A), 87:225–229.

Zygmund, A. (1959). *Trigonometric Series*, 2 vols., Cambridge University Press, Cambridge.

Notation

Index